Lecture Notes in Mathematics

Edited by A. Dold and B. Eckmann

1226

Alexandru Buium

Differential Function Fields and Moduli of Algebraic Varieties

Springer-Verlag
Berlin Heidelberg New York London Paris Tokyo

Author

Alexandru Buium
Department of Mathematics, National Institute for Scientific and Technical Creatior
B-dul Păcii 220, 79622 Bucharest, Romania

Mathematics Subject Classification (1980): 13N05, 14D20, 14L30

ISBN 3-540-17194-0 Springer-Verlag Berlin Heidelberg New York
ISBN 0-387-17194-0 Springer-Verlag New York Berlin Heidelberg

Printed in Germany

Printing and binding: Druckhaus Beltz, Hemsbach/Bergstr.
2146/3140-543210

INTRODUCTION.

Our background consists of two theories each having quite classical roots namely:

A) The theory of algebraic differential equations (ADE's) with no movable singularity and

B) The Galois theory of ADE's.

The first theory was initiated by Fuchs, Poincaré, Painlevé [Poin] [Pa] and has been given modern treatments through the work of several people (for a foliation-theoretic approach see Gérard-Sec [GS] and Jouanolou $[J_1]$ while for a differential algebraic approach in the one dimensional case see Matsuda [Mtd]). The second theory goes back to Picard and Vessiot and reached a very elegant and general form through the work of Kolchin $[Kol_n]$ $1 \leq n \leq 3$.

The primary goal of this research monograph is to relate the two theories above; this will turn out to be profitable for both of them.

To establish the link between A) and B) the first step is to develop a higher dimensional differential algebraic version of A). None of the methods used in [GS], $[J_1]$, [Mtd] seems suitable for this purpose: [GS] and $[J_1]$ are too "analytic" while [Mtd] is too related to the one-dimensional case. Our approach will be quite different and will lead us beyond our "primary goal", to what we called a "differential descent theory". This theory has an interest in itself and should be viewed as an "infinitesimal" analog of Shimura-Matsusaka theory of fields of moduli $[Sh_2]$, [Mtk]. Our proofs in this step will be combinations of moduli-theoretic methods (deformations of polarized algebraic varieties and compact analytic spaces) and differential algebraic methods (logarithmic derivatives on algebraic groups).

The second step in our approach will be Galois-theoretic. We shall use results proved in the first step plus Kolchin's differential Galois theory to describe in detail the interaction between A) and B). Proofs will also involve an analysis of K'/K-forms of quasi-homogenous projective varieties and some geometry of automorphisms of surfaces and abelian varieties.

The book is organized as follows.

In Chapter I we introduce our main objects and review some definitions and basic facts from differential algebra. A certain familiarity with the material in [Kol_1] and [Mtd] would be preferable but is not indispensable. An account of Kolchin's Galois theory is included.

Chapters II and III are new; they deal with the first and second steps described above respectively.

In Chapter IV we discuss the link between our theory and the classical analytic setting. Most facts presented in this Chapter are "well known to the experts" but there seems to be no suitable reference for them.

Internal references will be given by (X,y,z) or just (X,y) where X is the number of the chapter and y is the number of the paragraph; Within the same chapter we shall sometimes write (y,z) instead of (X,y,z).

Now we would like to explain in some detail our main applications; for simplicity we shall restrict ourselves to the "analytic case". So start with a region R in $\mathbb{C}^m$, let $w_1,\dots,w_m$ be coordinates in $\mathbb{C}^m$, put $\delta_j=\partial/\partial w_j$, consider the field of all meromorphic functions on R and $K\subset F$ subfields of it containing $\mathbb{C}$ such that K is relatively algebraically closed in F, F is finitely generated over K and $\delta_j(K)\subset K$, $\delta_j(F)\subset F$ for $1\leq j\leq m$. Denote by $\mathrm{Gal}_\Delta(F/K)$ the group of all K-automorphisms of F which commute with $\delta_1,\dots,\delta_m$ and call it the Δ-Galois group of F/K. We shall mainly be interested here in the following three properties:

(WN) F/K is called weakly normal if $F^{Gal_\Delta(F/K)}=K$.

(SN) F/K is called strongly normal if there exists a connected algebraic group G over $\mathbb{C}$ and a principal homogenous space W/K for G such that W is a model for F/K and the action of G on W induces an isomorphism $Gal_\Delta(F/K) \simeq G(\mathbb{C})$ (=set of $\mathbb{C}$-points of G).

(NMS) F/K is said to have no movable singularity if it has a projective model V such that $\delta_j(\mathcal{O}_V) \subset \mathcal{O}_V$ for $1 \leq j \leq m$.

The first definition is due to Kolchin [Kol_2] and is the first (and weakest) concept of normality one could think of but not much could be proved about it in general (cf. [Kol_2]). The second concept is also due to Kolchin [Kol_1], [Kol_2] (cf. also Białynicki-Birula [BB]); Kolchin's definition is in fact different from the one given above and the equivalence between the two definitions is a non-trivial fact (cf. [BB] or [Kol_1] p.430). One should say that strong normality has classical roots going back to Ehresmann's connections in principal bundles [NW]. It is related, as Kolchin's theory [Kol_3] shows, to the problem of "linearizing" algebraic differential equations by means of abelian functions. A lot of beautiful properties could be proved for strongly normal extensions (cf. [Kol_n] $1 \leq n \leq 3$): a "Galois correspondence" holds for such extensions and moreover strongly normal extensions with commutative group can be described explicitely in terms of "special values" of certain automorphic functions as it happens in classical class field theory, see (IV.1). The third definition is inspired from Matsuda's book [Mtd] where the case m=tr.deg.F/K=1 was treated; it has however classical roots too going back essentially to Fuchs and Poincaré [Poin]. We should emphasize that in definition of (NMS) δ_j cannot be interpreted as vector fields on V since they do not vanish on K (except of course the case $K=\mathbb{C}$ which corresponds to the case of differential equations with constant coefficients: this will appear in our setting as the trivial case !).

One of our main results will be the following:

THEOREM (III.3.1) (SN) is equivalent to (WN) + (NMS)

Using this "geometric characterisation" of strong normality we shall prove:

THEOREM (III.4.1) (SN) is equivalent to (WN) in each of the following cases:

1) tr.deg.F/K=1 ("curve" case)

2) tr.deg.F/K=2 and $\varkappa(F/K)\geqslant 0$ ("non-ruled surface" case) and

3) tr.deg.F/K=q(F/K) and $\varkappa(F/K)\geqslant 0$ (essentially the case of abelian varieties). Here $\varkappa$=Kodaira dimension and q=irregularity.

Note that case 1) for genus 0 is due to Kolchin [Kol_2] and was one of the starting points of our investigation; note also that condition $\varkappa(F/K)\geqslant 0$ in 2) cannot be removed as shown by an example of Kolchin (I.3.5).

Theorems above help one to get a better understanding of strong normailty. One the other hand one can prove:

THEOREM (III.2.1) If F/K has (NMS) then there exists an extension E/F such that E/K' is (SN) where K' is the algebraic closure of K in E.

Since strongly normal extensions have by Kolchin's theory an explicit description in terms of abelian functions and solutions of linear differential equations [Kol_3](see also (IV.1)) we are led to a differential algebraic solution of Poincaré's problem [Poin] of describing the "new transcendental functions" which may appear by integrating (systems of higher order) ADE's with "no movable singularity". We would like to note that the point of view of differential algebra is here much more precise than the classical point of view of analytic foliations, as explained in (IV.2).

The author wishes to express his sincere gratitude to Professor N.Radu for introducing him to differential algebra and to L.Bădescu and C.Bănică for continous encouragement and helpful discussions.

February 18, 1986 Alexandru Buium

CONTENTS

CHAPTER I. PRELIMINARIES

1. Terminology. Basic objects.

(1.1) Throughout the paper rings will be assumed commutative with 1-element and containing the field $\mathbb{Q}$ of rationals. All schemes will be over $\mathbb{Q}$.

If A is an integral domain $Q(A)$ will denote its field of quotients; if X is an integral scheme $Q(X)$ will denote its field of rational functions.

By a variety V over a field K we will always understand a quasi-projective geometrically integral scheme over K; if $p \in V$ the residue field at p will be denoted by $K(p)$.

If $A \longrightarrow B$ is a ring homomorphism and M is a B-module we denote as usual by $\mathrm{Der}_A(B,M)$ the B-module of A-derivations of B into M. If $A=\mathbb{Q}$ the subscript A will be omitted. We will also write $\mathrm{Der}_A(B)$ instead of $\mathrm{Der}_A(B,B)$ and $\mathrm{Der}(B)$ instead of $\mathrm{Der}(B,B)$. As well known the functor $\mathrm{Der}_A(B,-)$ is representable by $\Omega_{B/A}$ = module of differentials [Ha] p.172. Now if $X \longrightarrow S$ is a morphism of schemes and $\mathcal{F}$ is a quasi-coherent $\mathcal{O}_X$-module we denote as usual by $\Omega_{X/S}$ the $\mathcal{O}_X$-module of relative differentials [Ha] p.175 and put

$$\underline{\mathrm{Der}}_{\mathcal{O}_S}(\mathcal{O}_X,\mathcal{F}) := \underline{\mathrm{Hom}}_{\mathcal{O}_X}(\Omega_{X/S},\mathcal{F})$$

$$\mathrm{Der}_S(\mathcal{O}_X,\mathcal{F}) := \mathrm{Hom}_X(\Omega_{X/S},\mathcal{F}) = H^{\bullet}(X,\underline{\mathrm{Der}}_{\mathcal{O}_S}(\mathcal{O}_X,\mathcal{F}))$$

Remark that if $S=\mathrm{Spec}\,A$, $U=\mathrm{Spec}\,B$ is an open subset of X and $\mathcal{F}|_U = \tilde{M}$ for some B-module M then

$$H^{\bullet}(U,\underline{\mathrm{Der}}_{\mathcal{O}_S}(\mathcal{O}_X,\mathcal{F})) = \mathrm{Der}_A(B,M)$$

We shall write $\underline{\mathrm{Der}}_{\mathcal{O}_S}(\mathcal{O}_X)$ instead of $\underline{\mathrm{Der}}_{\mathcal{O}_S}(\mathcal{O}_X,\mathcal{O}_X)$ and $\mathrm{Der}_S(\mathcal{O}_X)$ instead of $\mathrm{Der}_S(\mathcal{O}_X,\mathcal{O}_X)$; note that $\underline{\mathrm{Der}}_{\mathcal{O}_S}(\mathcal{O}_X)$ is not in general a quasi-coherent sheaf because $\Omega_{X/S}$ is not in general coherent. In our applications this situation will often occur. If $S=\mathrm{Spec}\,\mathbb{Q}$ the subscripts $\mathcal{O}_S$ and S will be omitted. If $S=\mathrm{Spec}\,K$ with K a field and V is a variety over K then elements of $\mathrm{Der}_K(\mathcal{O}_V)$ will be called (global) vector fields on V.

We shall several times deal with a basic well known short exact sequence which we now recall. Let

$$X \xrightarrow{f} S=\mathrm{Spec}\,A \longrightarrow T=\mathrm{Spec}\,k$$

be morphisms of integral schemes with f dominant, k a field and $\Omega_{A/k}$ a flat A-module. Then there is an exact sequence

$$0 \longrightarrow f^*\Omega_{S/T} \xrightarrow{u} \Omega_{X/T} \longrightarrow \Omega_{X/S} \longrightarrow 0$$

Indeed we have to prove that u is injective; we may suppose $X=\mathrm{Spec}\,B$. Then we have a commutative diagram

$$\begin{array}{ccc}
\Omega_{A/k}\otimes_A B & \xrightarrow{\quad u \quad} & \Omega_{B/k} \\
\cap & & \downarrow \\
\Omega_{A/k}\otimes_A Q(B) & & \\
\| & & \\
\Omega_{Q(A)/k}\otimes_{Q(A)} Q(B) & \xrightarrow{\quad v \quad} & \Omega_{Q(B)/k}
\end{array}$$

with v injective by separability and we are done. Note that applying $\mathrm{Hom}_X(-,\mathcal{F})$ to the above exact sequence ($\mathcal{F}$ being quasi-coherent on X) we get an exact sequence of A-modules

$$\mathrm{Der}_T(\mathcal{O}_X,\mathcal{F}) \longrightarrow \mathrm{Der}_T(\mathcal{O}_S,f_*\mathcal{F}) \xrightarrow{\rho} \mathrm{Ext}^1(\Omega_{X/S},\mathcal{F})$$

The map ρ will be called the Kodaira-Spencer map associated to X,f S,T and $\mathcal{F}$.

(1.2) Now start with an arbitrary set Δ which we call the set of differential operators. Unlike in Kolchin's book [Kol$_1$] we do not suppose that Δ is finite; this is because some of our main applications (II.1.3),(II.3.8),(II.3.10) will involve infinite sets of (noncommuting) derivations. By a Δ-ring we mean a ring A together with a map

$$\begin{aligned} \Delta &\longrightarrow \mathrm{Der}(A) \\ \delta &\longmapsto \delta_A \end{aligned}$$

When there is no danger of confusion we write δa instead of $\delta_A a$ for $a \in A$, $\delta \in \Delta$. Define the ring of constants

$$A^{\Delta} = \{a \in A;\ \delta a = 0 \quad \text{for all} \quad \delta \in \Delta\}$$

We say that A is a partial Δ-ring if Δ is finite and $[\delta_A, \delta'_A] = 0$ for all $\delta, \delta' \in \Delta$ where $[\,,\,]$ denotes the Poisson bracket on $\mathrm{Der}(A)$. A will be called an ordinary Δ-ring if Δ is reduced to one element. An ideal I in A is called a Δ-ideal if $\delta(I) \subset I$ for all $\delta \in \Delta$. By a morphism of Δ-rings (or simply a Δ-morphism) we mean a ring homomorphism $f: A \longrightarrow B$ between Δ-rings such that $f(\delta a) = \delta(f(a))$ for all $a \in A$ and $\delta \in \Delta$. When A,B are fields we say that A,B are Δ-fields, $A \subset B$ (or B/A) is a Δ-field extension, or an isomorphism of A into B or that A is a Δ-subfield of B. Here are some basic facts about Δ-field extensions:

a) If F/K is an algebraic field extension then any derivation on K uniquely extends to F hence if K is a Δ-field there is a unique structure of Δ-field on F such that F/K is a Δ-field extension. Moreover F^{Δ}/K^{Δ} is easily seen to be also algebraic.

b) If K is a Δ-field then K^{Δ} is algebraically closed in K.

c) If F/K is a Δ-field extension then F^{Δ} and K are linearly disjoint over K^{Δ} (for a proof see [BB] p.93).

Suppose now $f_1: A \longrightarrow B_1$ and $f_2: A \longrightarrow B_2$ are Δ-morphisms; then there is a structure of Δ-ring on $B_1 \otimes_A B_2$ making $B_1 \otimes_A B_2$ the fibred sum of B_1 and B_2 in the category of Δ-rings; it is given by the formula $\delta(x\otimes y)=x\otimes\delta y + \delta x\otimes y$. The following particular case will often appear: suppose K is a Δ-field, C is a subfield of K^{Δ} and R is a C-algebra. Then for any $\delta\in\Delta$ there is a unique derivation δ_K^* on $R\otimes_C K$ satisfying $\delta_K^*(x\otimes y)=x\otimes\delta y$ for all $x\in R$ and $y\in K$; it will be called the trivial lifting of δ_K to $R\otimes_C K$.

Finally note that if A is a Δ-ring and S is a multiplicative system in A then there is a unique structure of Δ-ring on $S^{-1}A$ making $A\longrightarrow S^{-1}A$ a Δ-morphism.

Now one can make similar definitions for schemes instead of rings. So by a Δ-scheme we mean a scheme V together with a map

$$\Delta\longrightarrow \mathrm{Der}(\mathcal{O}_V)$$
$$\delta\longmapsto \delta_V$$

It will be called a partial Δ-scheme if Δ is finite and the δ_V's are pairwise commuting. Analog definition for ordinary Δ-scheme. A point $p\in V$ will be called a Δ-point if $\delta(m_p)\subset m_p$ where m_p is the maximal ideal of the local ring $\mathcal{O}_{V,p}$. We shall denote by V_Δ the set of all Δ-points of V. A closed subscheme W of V will be called a Δ-subscheme if $\delta(I_W)\subset I_W$ where I_W is the ideal sheaf of W. Note that any Δ-subscheme has a natural structure of Δ-scheme. A morphism of Δ-schemes $f:V\longrightarrow W$ (or simply a Δ-morphism) will mean a morphism of schemes such that the corresponding homomorphism $\mathcal{O}_W\longrightarrow f_*\mathcal{O}_V$ induces morphisms of Δ-rings. By a Δ-variety V over a Δ-field K we will mean a morphism of Δ-schemes $f:V\longrightarrow \mathrm{Spec}\,K$ such that V is a variety over K. One of the main features of the theory is that for a Δ-variety V the derivations δ_V are not vector fields on V (since they do not vanish on K); see also (IV.2). If V,W are Δ-varieties over K and u is a rational map from V to W we say that u is a Δ-rational map if the morphism $V_0\longrightarrow W$ (V_0=locus where u is defined) is a Δ-morphism. Finally exactly as in (1.1) fibre products exist in the category of Δ-schemes. In particular if K is a Δ-field, C is a subfield of K^{Δ} and X is a variety over C then there exists a unique structure of Δ-variety of $X\otimes_C K$ over K for which $Q(X)\otimes 1 \subset Q(X\otimes_C K)^{\Delta}$; derivations for this structure will be denoted by δ_K^* (so $\delta_K^*(x\otimes y)=x\otimes\delta y$ for all $\delta\in\Delta$, $x\in Q(X)$, $y\in K$) and will be called the trivial liftings of δ_K to $X\otimes_C K$.

Let $K\subset F$ be a Δ-field extension. It will be called a Δ-function field if K is relatively algebraically closed in F and F is finitely generated over K as a field extension. By a model of F/K we understand as usual a variety V over K whose function field $Q(V)$ is K-isomorphic to F; as a rule we shall identify F with $Q(V)$. A model V of F/K will be called a Δ-model if $\delta(\mathcal{O}_V)\subset\mathcal{O}_V$ for all $\delta\in\Delta$ (hence such a V is in a natural way a Δ-variety over K). Our basic definition below is inspired from Matsuda's work [Mtd]; terminology will be explained in (IV.2).

(1.3) Definition. A Δ-function field F/K is said to have no movable singularity if it has a projective Δ-model.

As we shall see below (2.3) if Δ is finite than any Δ-function field has a Δ-model so the key word in the above definition is "projective". Another important concept will be the following:

(1.4) Definition. A Δ-function field is said to be split if it is generated by constants i.e. $F=K(F^{\Delta})$.

Split Δ-function fields with $tr.deg.F/K=1$ were called in [Mtd] Clairaut extensions; this name was motivated by their connection with the Clairaut equation. Our terminology is explained by (1.6) below. The interplay between split extensions and "no movable singularity" in the case $tr.deg.F/K=1$ was one of the main ideas in [Mtd] but the proofs heavily relied on "curve techniques". As we shall see this interplay will survive in the case $tr.deg.F/K\geqslant 2$ too but due to entirely different arguments.

The following definition is inspired from [NW] p.957:

(1.5) Definition. Let K be a Δ-field with constant field C and V a Δ-variety over K. We say that V is split if there exists a variety X over C and a K-isomorphism $\sigma : V \longrightarrow X \otimes_C K$ such that

$$(\sigma^*)^{-1} \delta_V \sigma^* = \delta_K^* \qquad \text{for all } \delta \in \Delta$$

Such a σ will be called a splitting isomorphism. Here σ^* is of course the induced K-isomorphism $Q(X \otimes_C K) \longrightarrow Q(V)$ and δ_K^* is the trivial lifting of δ_K to $X \otimes_C K$. We will also say that that V is a split Δ-model of $Q(V)/K$.

(1.6) LEMMA. Let F/K be a Δ-function field. Then F/K is split if and only if it has a split projective Δ-model.

Proof. To check the "if" part give $X \otimes_C K$ the structure of a Δ-variety by putting

$$\delta_{X \otimes_C K} = \delta_K^* \qquad \text{for all } \delta \in \Delta$$

Then the Δ-varieties $X \otimes_C K$ and V will be K-isomorphic hence so will be the Δ-function fields $Q(X \otimes_C K)/K$ and F/K; but now the former is generated by $Q(X) \otimes 1 \subset Q(X \otimes_C K)^{\Delta}$ and we are done. To prove the "only if" part note that it follows from [BB] p.94 that F^{Δ} is finitely generated over $C = K^{\Delta}$. On the other hand F^{Δ} and K are linearly disjoint over C (1.2) hence $F = Q(F \otimes_C K)$. Now take any projective model X of F^{Δ}/C; then $X \otimes_C K$ will be a split projective Δ-model of F/K and we are done.

We obtained in this way a first (and in some sense trivial) example of Δ-function field with no movable singularity:

(1.7) COROLLARY. If F/K is a split Δ-function field then it has no movable singularity.

A second example (of a quite different nature) will be given in (III.3.1) where we shall prove that a Δ-function field which is strongly normal in Kolchin's sense -see (3.1) below - has no movable singularity.

One moral of our theory will be roughly speaking that Δ-function fields with no movable singularity can be reduced in some sense to the two examples we mentioned cf. (III.1) and (III.2). Another moral will be that any Δ-function field with no movable singularity "splits" over some base change (II.2.2).

(1.8) We would like to close this paragraph by calling the attention on the fact that our Δ-schemes are beings quite different from the differential schemes appearing in [Kei] or from the Ritt schemes appearing in [Bu_4]. The relation between them is the following: for any Δ-scheme V the set V_Δ may be given a natural structure of ringed space by taking on V_Δ the induced topology from V and putting $\mathcal{O}_{V_\Delta} = j^{-1}(\mathcal{O}_V)$ where $j: V_\Delta \to V$ is the inclusion. The ringed space V_Δ just defined is a differential scheme in the sense of [Kei] and a Ritt scheme in the sense of [Bu_4]. It will play no role in our approach. On the contrary our computations of the set V_Δ in (II.3) can be used to provide many interesting examples of Ritt schemes.

2. Geometry of Δ-varieties.

In this § we summarize some geometric properties of Δ-varieties. Most results are essentially well known but references are available sometimes only for the case of Δ-varieties over a constant Δ-field (i.e. for varieties equiped with a set of vector fields). It will be convenient for us to give below a quick treatment of the general case.

(2.1) LEMMA. Let A be a noetherian ring and $\delta \in Der(A)$. Then

1) $\delta(nil(A)) \subset nil(A)$ where $nil(A)$=nilpotent radical of A,

2) $\delta(P) \subset P$ for any minimal prime P in Spec A

3) If A is integral and we still denote by δ the induced derivation on $Q(A)$ then $\delta(A^{nor}) \subset A^{nor}$ where A^{nor} is the normalisation of A.

4) If A is finitely generated over some field and $P \in$ Spec A is such that A_P is not regular but A_Q is regular for any prime Q contained in P and different from P, then $\delta(P) \subset P$.

Proof. 1) is [Ka] Lemma 1.8; 2) is proved for instance in [Ra]; 3) is a theorem of Seidenberg [Se_1]; 4) is also essentially due to Seidenberg [Se_2] (for a proof of this form see [Mtm] p.233).

(2.2) PROPOSITION. Let V be a noetherian Δ-scheme. Then

1) V_{red} has a natural structure of Δ-scheme,

2) The irreducible components of V (with their reduced structure) are Δ-subschemes of V,

3) If V is integral then its normalisation V^{nor} has a natural structure of Δ-scheme,

4) If V is a Δ-variety over some Δ-field then all irreducible components of the singular locus V_{sing} are Δ-subschemes of V.

5) If W is an integral subscheme of V and $p \in V$ is the generic

point of W then $p \in V_\Delta$ if and only if W is a Δ-subscheme of V.

6) If $W_1, \dots, W_s$ are Δ-subschemes of V so is $W_1 \cap \dots \cap W_s$.

7) If V is integral and normal and $x \in Q(V)^\Delta$ then all irreducible components of the principal divisor $\mathrm{div}(x)$ are Δ-subschemes of V.

8) If V is a Δ-variety over some Δ-field K and if p is a maximal element in V_Δ then the extension $K(p)^\Delta/K^\Delta$ is algebraic.

Proof. 1)-4) follow from the corresponding assertions in (2.1); 5) is trivial; 6) is an effect of the fact that the sum of a family of Δ-ideals is also a Δ-ideal; 7) is a consequence of 2) in (2.1) applied to $\mathcal{O}_P/x\mathcal{O}_P$ or $\mathcal{O}_P/x^{-1}\mathcal{O}_P$ where $P \in V$ has codimension one and of 5) above. Now 8) in the "partial case" is a consequence of a result due to Kolchin [Mtd] p.108 but the proof there really depends on the hypothesis that one is dealing with partial Δ-rings (since it makes use of ellimination theory from differential algebra). We shall prove 8) by making use of the usual ellimination theory from algebra (i.e. of Chevalley's constructibility theorem). Let $\mathrm{Spec}\, A$ be an affine open neighbourhood of p in V. Then p will be a maximal element of $(\mathrm{Spec}\, A)_\Delta$. Suppose $Q(A/p)^\Delta$ contains a transcendental element t over $C = K^\Delta$. We have a rational map from $\mathrm{Spec}(A/p)$ to $\mathrm{Spec}\, K[t]$ hence a morphism $u : \mathrm{Spec}(A_f/pA_f) \longrightarrow \mathrm{Spec}\, K[t]$ for some $f \in A \setminus p$. By Chevalley's theorem there is an element $g \in K[t]$ different from zero such that $\mathrm{Spec}(K[t]_g) \subset \mathrm{Im}(u)$. Choose any $a \in C$ such that $g(a) \neq 0$. Then $(t-a)K[t]$ is a Δ-ideal in $K[t]$ contained in $\mathrm{Im}(u)$ so the "fibre" $(A_f/pA_f)/(t-a)$ is a non-zero Δ-ring hence by 2) in (2.1) it has at least a prime Δ-ideal. The inverse image of this ideal in A will be a prime Δ-ideal containing p and different from p, contradiction. The proposition is proved.

(2.3) PROPOSITION. Let F/K be a Δ-function field with Δ finite and let V be a model of F/K. Put

$$V_o = \{p \in V;\ \delta(\mathcal{O}_{V,p}) \subset \mathcal{O}_{V,p} \text{ for all } \delta \in \Delta\}$$

Then V_o is Zariski open in V. Moreover if V is normal then $V \setminus V_o$ has pure codimension one. Finally if V is regular the set of effective divisors D on V for which $\delta_V \in \mathrm{Der}(\mathcal{O}_V, \mathcal{O}_V(D))$ for all $\delta \in \Delta$ has a smallest element (which we call the divisor of movable singularities on V). Its support is $V \setminus V_o$.

Proof. V_o is open by $[Bu_1]$ p.58. For convenience we reproduce the argument. Let $\mathrm{Spec}\, A$ be an open set of V containing some $p \in V_o$ with A generated over K by $a_1, \ldots, a_N \in A$. Put $a_i = f_i(\delta)/g_i(\delta)$ for all $\delta \in \Delta$ with $f_i(\delta), g_i(\delta) \in A$ and put $g = \prod g_i(\delta)$ where the product is taken after all i and δ. Then clearly $\delta(A_g) \subset A_g$ for all $\delta \in \Delta$ hence $\mathrm{Spec}\, A_g \subset V_o$ and we are done.

Now if V is normal and p is the generic point of some irreducible component of $V \setminus V_o$ then since $\mathcal{O}_p = \cap \mathcal{O}_q$ the intersection being taken after all codimension one points q of whom p is a specialisation it follows that some q must belong to $V \setminus V_o$ hence $p = q$. Finally if V is regular the existence of the divisor of movable singularities follows immediately from the formula

$$\mathcal{O}_V(D_1) \cap \mathcal{O}_V(D_2) = \mathcal{O}_V(D_3)$$

where D_1, D_2 are effective divisors, $D_3 = \inf(D_1, D_2)$ and all $\mathcal{O}_V(D_i)$ are viewed as subsheaves of the constant sheaf $Q(V)$.

We should remark that V_o defined in the proposition above is a Δ-model of F/K so V itself is a Δ-model if and only if the divisor of movable singularities (in case V is regular) vanishes. As we shall see in (III.4) the geometry of the divisor of movable singularities will play a role in our Galois-theoretic results.

(2.4) PROPOSITION. Let $f:V\longrightarrow W$ be a morphism of Δ-schemes. Then the following hold:

1) $f(V_\Delta)\subset W_\Delta$.

2) For any Δ-subscheme W_1 of W the scheme $f^{-1}(W_1)$ is in a canonical way a Δ-scheme.

Proof. Obvious.

For the propositions (2.5) and (2.6) below compare with [Li_1] where the case of "vector fields" is treated.

(2.5) PROPOSITION. Let $f:V\longrightarrow W$ be a dominant morphism of integral noetherian schemes and suppose V has a structure of Δ-scheme.

1) If f is flat and surjective and if $Q(W)$ is a Δ-subfield of $Q(V)$ then W has a natural structure of Δ-scheme (here "natural" means that δ_W are induced by $\delta_{Q(W)}$).

2) If $f_*\mathcal{O}_V=\mathcal{O}_W$ then W has a natural structure of Δ-scheme (again δ_W being induced by δ_V so f will become a Δ-morphism).

3) If f is a Δ-morphism and if $q\in V_{reg}$, $p:=f(q)\in W_{reg}$, p has codimension one in W and $q\notin(f^{-1}(p))_{reg}$ then $p\in W_\Delta$ (roughly speaking this says that at least when V is regular and W is normal the codimension one components of the "discriminant" of f are Δ-subschemes of W).

Proof. 1) For any $p\in V$ flatness yelds $\mathcal{O}_{W,f(p)}=\mathcal{O}_{V,p}\cap Q(W)$ hence $\delta(\mathcal{O}_{W,f(p)})\subset\mathcal{O}_{W,f(p)}$ for all $\delta\in\Delta$.

2) Consider the canonical map $f^*\Omega_{W/\mathbb{Q}}\longrightarrow\Omega_{V/\mathbb{Q}}$. Taking $Hom_V(-,\mathcal{O}_V)$ we get a natural "restriction" map

$$Der(\mathcal{O}_V)\longrightarrow Hom_V(f^*\Omega_{W/\mathbb{Q}},\mathcal{O}_V)=Hom_W(\Omega_{W/\mathbb{Q}},f_*\mathcal{O}_V)=Der(\mathcal{O}_W)$$

3) Let t be a parameter of $\mathcal{O}_{W,p}$. Since $\mathcal{O}_{V,q}$ is regular but $\mathcal{O}_{V,q}/t\mathcal{O}_{V,q}$ is not we get $t\in m_q^{\,2}$ hence for all $\delta\in\Delta$

$$\delta t \in \delta(m_q^2) \cap \mathcal{O}_{W,p} \subset m_q \cap \mathcal{O}_{W,p} = m_p$$

hence $\delta(m_p) \subset m_p$ and we are done.

(2.6) PROPOSITION. Let V be an integral noetherian Δ-scheme and Z a subscheme. Let $W \longrightarrow V$ be the blowing up of V with respect to the ideal sheaf I_Z of Z.

1) If Z is a Δ-subscheme then W has a natural structure of Δ-scheme (δ_W being induced by $\delta_{Q(V)}$)

2) If W is a Δ-scheme (with δ_W induced by $\delta_{Q(V)}$) and if Z is integral of codimension ≥ 2 in V and V is regular at the generic point of Z then Z is a Δ-subscheme of V.

Proof. 1) We may suppose V=Spec A, Z=Spec A/I, $I=(x_1,\ldots,x_t)$. W will be covered by open sets Spec B_k where

$$B_k = A\left[x_1/x_k,\ldots,x_t/x_k\right] \subset Q(A)$$

By hypothesis we have

$$\delta x_i = \sum_{s=1}^{t} a_{is}(\delta)x_s \qquad \text{with} \quad a_{is}(\delta) \in A$$

we get

$$\delta(x_i/x_k) = \sum_{s=1}^{t} (a_{is}(\delta)(x_s/x_k) - a_{ks}(\delta)(x_s/x_k)(x_i/x_k)) \in B_k$$

so $\delta(B_k) \subset B_k$ and we are done.

2) Put $T=\mathrm{Spec}(\mathcal{O}_{V,p})$ where p is the generic point of Z. Then $W \times_V T$ is the blowing up of T at its closed point m_p. Let $y_1 \ldots$ $\ldots, y_s$ be a regular system of parameters for $\mathcal{O}_{V,p}$. By (1.2) $W \times_V T$ has a natural structure of Δ-scheme and for any $\delta \in \Delta$

$$(\delta y_j)/y_k - (y_j \delta y_k)/y_k^2 = \delta(y_j/y_k) = F(y_1/y_k,\ldots,y_s/y_k)$$

where F is a polynomial with coefficients in $\mathcal{O}_{V,p}$. Multiplying the relation above by a suitable power of y_k and using the fact that the graded ring of $\mathcal{O}_{V,p}$ with respect to its maximal ideal m_p is a polynomial ring over its residue field in the indeterminates $\hat{y}_1,\ldots,\hat{y}_s$ where $\hat{y}_i=y_i \bmod m_p^2$ we get that $\delta y_i \bmod m_p=0$ hence $\delta(m_p)\subset m_p$ and we are done.

(2.7) PROPOSITION. Let $f:W\longrightarrow V$ be an étale map of integral schemes and suppose V is a Δ-scheme. Then W is a Δ-scheme in a unique way such that f is a Δ-morphism.

Proof. By [Mi] p.26 for any $y\in W$ there are open neighbourhoods Spec B and Spec A of y and $f(y)$ respectively such that f(Spec B)$\subset$Spec A and $B=(A[T]/(F))_b$ with $F\in A[T]$ monic, $b\in A[T]/(F)$ and dF/dT a unit in B. Let t be the image of T in B. From $F(t)=0$ we get $F^{\delta}(t)+(dF/dT)\,\delta t=0$ where F^{δ} is obtained from F by applying δ to each coefficient. Since $\delta(A)\subset A$ we get $\delta t\in B$ hence $\delta(B)\subset B$ and we are done. Here δt meant of course $\delta_{Q(B)}t$.

(2.8) PROPOSITION. Let V be a Δ-scheme, $L\in \mathrm{Pic}(V)$, $s\in H^0(V,L^k)$ and $W=\mathrm{Spec}(\mathcal{O}_V\oplus L^{-1}\oplus\ldots\oplus L^{-k+1})$ be the cyclic covering defined by s. Suppose that the zero subscheme defined by s in V is a Δ-subscheme. Suppose also that both V and W are integral. Then there is a unique structure of Δ-scheme on W making $W\longrightarrow V$ a Δ-morphism.

Proof. The problem is local so we may suppose that $V=\mathrm{Spec}\,A$, $s\in A$ $W=\mathrm{Spec}\,B$, $B=A[t]=A[T]/(T^k-s)$. By hypothesis $\delta s=a(\delta)s$ for all $\delta\in\Delta$ where $a(\delta)\in A$. Now define $\delta_{A[T]}$ by $\delta T=k^{-1}a(\delta)T$. One immediately checks that $\delta(T^k-s)\in(T^k-s)A[T]$ hence $\delta_{A[T]}$ induce derivations δ_B and since $\delta_{Q(A)}$ uniquely extend to $Q(B)$ we are done.

We close this § by discussing some results of Burns and Wahl cf. [BW], [Wa]. By surface we shall always mean here a variety of dimension 2 over an algebraically closed field. By a curve on a surface we will understand any Cartier divisor. If V is a non-singular projective surface and E is a curve on V we say that E is exceptional if there is a morphism $f:V \longrightarrow W$ onto a projective normal surface W such that $f(E)$ is a finite set and $V\setminus E \longrightarrow W\setminus f(E)$ is an isomorphism. We say E is minimal if it does not contain exceptional curves of the first kind in its support. The following result is essentially contained in [BW] where the case of vector fields was treated:

(2.9) PROPOSITION. Let F/K be a Δ-function field (Δ finite) of transcendence degree 2 with K algebraically closed and V a non-singular projective model of F/K. Let D be the divisor of movable singularities on V and let E be a minimal exceptional divisor on V. Suppose $\mathrm{Supp}\, D \subset E$. Then $D=0$ i.e. V is a Δ-model.

Proof. In [BW] a proof was given in the case $K=K^{\Delta}$. The argument immediately extends to the general case. We recall it here for convenience. Let $f:V \longrightarrow W$ be a "contraction" of E as in the definition above. As well known there exists a sequence

$$W_n \xrightarrow{f_n} W_{n-1} \xrightarrow{f_{n-1}} \cdots \xrightarrow{f_2} W_1 \xrightarrow{f_1} W_o=W$$

where W_n is non-singular and f_k is the blowing up of the (reduced) ideal of the singular locus of $(W_{k-1})^{nor}$. By 3) and 4) in (2.2) and by (2.6) W_n is a Δ-model of F/K. Since $V \longrightarrow W$ is the minimal desingularisation of W the rational map $W_n \dashrightarrow V$ is everywhere defined hence by 2) in (2.5) V is a Δ-model and we are done.

The proof above shows that any Δ-function field F/K with no movable singularity with $\mathrm{tr.deg.}F/K=2$ has a non-singular projective Δ-model provided K is algebraically closed. The same holds (for trivial reasons) if $\mathrm{tr.deg.}F/K=1$ (cf.3) in (2.2)). One can ask the question whether this holds for arbitrary transcendence degree. One knows this is so if $K^{\Delta}=K$ by a deep result of Hironaka [Li_1] p.106 on equivariant resolutions:

(2.10) PROPOSITION. Let X be a projective variety over an algebraically closed field (characteristic zero as usual !). Then there exists a birational morphism $\tilde{X} \longrightarrow X$ from a non-singular projective variety $\tilde{X}$ such that any vector field on X lifts to a (unique) vector field on $\tilde{X}$.

Using (2.10) and our "differential descent theory" in (II .1) we shall prove that indeed any Δ-function field F/K with no movable singularity has a non-singular projective Δ-model provided K is algebraically closed (II .1.24).

As we shall see the non-negativity behaviour of the divisor of movable singularities reflected by (2.9) will be used in an essential way in the proof of (III.4.1); it would be very interesting from this point of view to dispose of a higher dimensional analog of (2.9). The following result emerging directly from [Wa] gives a hint of what this generalisation should be:

(2.11) PROPOSITION. Let F/K be a Δ-function field (Δ finite) with K algebraically closed, let V be a non-singular projective model of F/K and let D be the divisor of movable singularities on V. Suppose $\operatorname{Supp} D = E$ is a non-singular subvariety of V with ample conormal bundle N_E^{-1}. Then there exists a nonsingular projective Δ-model W of F/K and a closed point $p \in W \setminus W_{\Delta}$ such that V is the blowing up of W at p and E is the exceptional divisor.

Proof. For any integer $m \geqslant 1$ let mE denote both the Cartier divisor and the closed subscheme of V with ideal sheaf $\mathcal{O}_V(-mE)$. Taking $\operatorname{Hom}_V(-,N_{mE})$ and $\operatorname{Hom}_V(\Omega_{V/\mathbb{Q}},-)$ in the exact sequences

$$0 \longrightarrow g^*\Omega_{K/\mathbb{Q}} \longrightarrow \Omega_{V/\mathbb{Q}} \longrightarrow \Omega_{V/K} \longrightarrow 0$$

$$0 \longrightarrow \mathcal{O}_V \longrightarrow \mathcal{O}_V(mE) \longrightarrow N_{mE} \longrightarrow 0$$

(where $g:V \longrightarrow \operatorname{Spec} K$) we obtain

(1) $$0 \rightarrow H^0(V, \underline{\mathrm{Der}}_K(\mathcal{O}_V)\ N_{mE}) \rightarrow \mathrm{Der}(\mathcal{O}_V, N_{mE}) \rightarrow \mathrm{Hom}_K(\Omega_{K/\mathbb{Q}}, H^0(N_{mE}))$$

(2) $$0 \longrightarrow \mathrm{Der}(\mathcal{O}_V) \longrightarrow \mathrm{Der}(\mathcal{O}_V, \mathcal{O}_V(mE)) \rightarrow \mathrm{Der}(\mathcal{O}_V, N_{mE})$$

Consider also the exact sequences ($k \geqslant 1$):

(3) $$0 \longrightarrow N_{kE} \longrightarrow N_{(k+1)E} \longrightarrow N_E^{k+1} \longrightarrow 0$$

Using (3) and anmpleness of N_E^{-1} we get $H^0(N_{mE})=0$. Three cases may occur:

Case 1): (E, N_E^{-1}) is different from both $(\mathbb{P}^N, \mathcal{O}_{\mathbb{P}^N}(1))$ and $(\mathbb{P}^1, \mathcal{O}_{\mathbb{P}^1}(2))$. In this case by Wahl's theorem [Wa] one has

$$H^0(E, \underline{\mathrm{Der}}_K(\mathcal{O}_E) \otimes N_E^k) = 0 \quad \text{for} \quad k \geqslant 1$$

and this implies, using the standard exact sequence

(4) $$0 \longrightarrow \underline{\mathrm{Der}}_K(\mathcal{O}_E) \longrightarrow \underline{\mathrm{Der}}_K(\mathcal{O}_V) \otimes \mathcal{O}_E \longrightarrow N_E \longrightarrow 0$$

that

$$H^0(V, \underline{\mathrm{Der}}_K(\mathcal{O}_V) \otimes N_E^k) = 0 \quad \text{for} \quad k \geqslant 1.$$

Using (3) again we get by induction that

$$H^0(V, \underline{\mathrm{Der}}_K(\mathcal{O}_V) \otimes N_{mE}) = 0.$$

We conclude by (1) and (2) that $\mathrm{Der}(\mathcal{O}_V) = \mathrm{Der}(\mathcal{O}_V, \mathcal{O}_V(mE))$ for all $m \geqslant 1$ contradicting the definition of D (2.3).

Case 2): $(E, N_E^{-1}) = (\mathbb{P}^1, \mathcal{O}_{\mathbb{P}^1}(2))$. This case is impossible by (2.9) plus Artin's contractibility criterion [A]

Case 3): $(E, N_E^{-1}) = (\mathbb{P}^N, \mathcal{O}_{\mathbb{P}^N}(1))$ leads to our conclusion (use (2.6)).

3. Kolchin's differential Galois theory.

In this § we review some basic facts from Kolchin's theory $[Kol_n]$ $1 \leq n \leq 4$ (cf. also Białynicki-Birula [BB]) and put them in a form suitable for our applications. We divide the § into three sections.

A. Galois correspondence.

(3.1) Let F/K be a Δ-field extension. Put $Gal_\Delta(F/K)$ = group of all Δ-automorphisms of F over K. The notation $Gal(F/K)$ will be reserved for the set of all K-automorphisms of F not necessarily commuting with the members of Δ. We call $Gal_\Delta(F/K)$ the Δ-Galois group of F/K. Following Kolchin $[Kol_2]$ we say that F/K is weakly normal if

a) F is a partial Δ-field finitely generated over K as a field extension.

b) $K^\Delta = F^\Delta$ and K^Δ is algebraically closed.

c) The field of invariants of $Gal_\Delta(F/K)$ in F is K in other words

$$F^{Gal_\Delta(F/K)} = K$$

Following Kolchin $[Kol_2]$ we say that F/K is strongly normal if conditions a),b) above hold and in addition we have

c') For any K-isomorphism σ of F into a partial Δ-field extension E of F both extensions $F \subset F\sigma F$ and $\sigma F \subset F\sigma F$ are split (here $F\sigma F$ denotes of course the compositum of F and σF in E). In fact as shown in $[Kol_1]$ p.393 the condition b) in the definition of strong normality can be weakened to "K^Δ is algebraically closed"; moreover Kolchin developed in $[Kol_1]$ a theory in which b) was completely removed; we shall not be concerned with this more general case.

The following is an immediate consequence of definitions (cf. $[Kol_1]$

p.393): let E/K be a Δ-field extension and let F_1, F_2 be intermediate Δ-fields. Suppose $E^{\Delta}=K^{\Delta}$ and both F_1/K and F_2/K are strongly normal; then F_1F_2/K is strongly normal.

(3.2) A useful remark concerning invariants of groups acting on function fields is the following (cf. [Ro] pp.405-406). Suppose $C\subset D$ is a field extension with C algebraically closed in D, let G be a group of C-automorphisms of D and let $C\subset K$ be any field extension. Of course G will act on $Q(D\otimes_C K)$ in a natural way by K-automorphisms. Then

$$(Q(D\otimes_C K))^G=Q(D^G\otimes_C K)$$

An immediate consequence of this fact is that if F/K is a weakly normal Δ-function field and if K_1/K is an algebraic extension then K_1F/K_1 is still weakly normal (here $K_1F=K_1\otimes_K F$ has a canonical structure of Δ-field by (1.2)). Indeed the extension $(K_1F)^{\Delta}/F^{\Delta}$ is algebraic so $(K_1F)^{\Delta}=(K_1)^{\Delta}=K^{\Delta}$ and since by base change one gets an embedding of $Gal_{\Delta}(F/K)$ into $Gal_{\Delta}(K_1F/K_1)$ one has

$$(F\otimes_K K_1)^{Gal_{\Delta}(K_1F/K_1)}\subset Q(F^{Gal_{\Delta}(F/K)}\otimes_K K_1)=K_1$$

(3.3)PROPOSITION [BB]. Let F/K be a partial Δ-function field and $C:=K^{\Delta}=F^{\Delta}$ algebraically closed. The following are equivalent:

1) F/K is strongly normal.

2) There exists a connected algebraic group G over C and a principal homogenous space W/K for G such that W is a model of F/K and $G(C)\subset Gal_{\Delta}(F/K)$.

Moreover the pair (G,W) in 2) is uniquely determined up to isomorphism and we have in fact $G(C)=Gal_{\Delta}(F/K)$.

Let's make some comments. In the above statement as well as througr out the paper by an algebraic group G over a field C (not necessarily algebraically closed) we will mean a group scheme of finite type over C. Such a G is always quasi-projective over C by Chow's theorem [Ch], [Ray]. By a principal homogenous space W/K for G (PHS) where G is connected and K is some field extension of C we understand (cf. [Mi] p.120) a scheme W of finite type over K plus an action $W\times_C G \longrightarrow W$ such that the induced morphism $W\times_C G \longrightarrow W\times_K W$ is an isomorphism. By [Ch], [Ray] again such a W is quasi-projective over K hence a variety. Clearly the group G(C) of C-points of G acts on W/K hence G(C) acts on Q(W) by K-automorphisms; the condition $G(C)\subset Gal_\Delta(F/K)$ in (3.3) says that these K-automorphisms should commute with all δ_F's. Due to the unicity of G in (3.3) we may write $G=G_{F/K}$.

(3.4) Here is now what Kolchin's theory [Kol_1] gives. For any strongly normal extension F/K (which is not necessarily a Δ-function field) one constructs in a "natural way" an algebraic group $G_{F/K}$ over $C:=K^\Delta=F^\Delta$ (not necessarily connected but which coincides with the group $G_{F/K}$ from (3.3) provided F/K is a Δ-function field) having the following properties:

1) $Gal_\Delta(F/K)$ "naturally" identifies with $G_{F/K}(C)$,

2) $\dim G_{F/K} = tr.deg.F/K$,

3) $G_{F/K}$ is connected if and only if F/K is a Δ-function field,

4) For any intermediate Δ-field K_1 between K and F the extension F/K_1 is strongly normal and G_{F/K_1} "naturally" identifies with an algebraic subgroup of $G_{F/K}$. Moreover $K_1 \dashrightarrow G_{F/K_1}$ gives the usual 1-1 correspondence between intermediate Δ-fields and algebraic subgroups of $G_{F/K}$. Finally the following are equivalent:

α) K_1/K is strongly normal

β) K_1/K is weakly normal

γ) G_{F/K_1} is normal in $G_{F/K}$

and if the above equivalent conditions hold then $G_{K_1/K}\simeq G_{F/K}/G_{F/K_1}$.

5) If E is any partial Δ-field extension of F with $E^\Delta=F^\Delta$

and K_1 is an intermediate Δ-field between K and E then K_1F/K_1 is strongly normal and G_{K_1F/K_1} "naturally" identifies with $G_{F/F\cap K_1}$.

What "naturality" means in the above statements will not be made explicit here; we send for this to Kolchin's book [Kol_1]. However we should note that we slightly shifted notations in our presentation and also we avoided the use of Kolchin's universal Δ-field; this is essentially because our algebraic groups are schemes rather than Weil-like beings as in Kolchin's book. It will be not difficult however to make our exposition agree with [Kol_1].

So far we have seen that strongly normal extensions have a "good" Galois theory; in particular by 4) above they are weakly normal. On the other hand weak normality of F/K does not imply strong normality as shown by an example of Kolchin [Kol_2]p.795. In Kolchin's example $K=K^{\Delta}=\mathbb{C}$ and tr.deg.$F/K=4$. It is of some interest for us (cf. (III.41)) to get examples with smaller transcendence degree by allowing K not to be algebraically closed or constant. And indeed this can be done by slightly modifying Kolchin's example:

(3.5) Claim. The ordinary Δ-field extension

$$K=\mathbb{C}(e^x,e^{ix})\subset\mathbb{C}(x,e^x,e^{ix},e^{x^2})=F$$

is weakly normal but not strongly normal. Here the two fields are viewed as Δ-subfields of $\mathbb{C}((x))$ with derivation d/dx.

To prove our claim note that weak normality was proved in [Kol_2] loc.cit. Now for any $d\in\mathbb{C}$ the formulae

$$\begin{aligned} x &\longmapsto x+d\\ e^x &\longmapsto e^x\\ e^{ix} &\longmapsto e^{ix}\\ e^{x^2} &\longmapsto e^{2dx}e^{x^2}\end{aligned}$$

define a K-isomorphism $\sigma:F\longrightarrow\mathbb{C}((x))$ (of Δ-fields). If F/K

was strongly normal we would get $\sigma F=F$ which fails for $2d \notin \mathbb{Z}[i]$.

Note that Białynicki-Birula developed in [BB] a Galois theory for not necessarily partial Δ-function fields. We will need the following corollary of his theory at some point in (III.2.2):

(3.6) PROPOSITION. Let C be an algebraically closed field, G a connected algebraic group over C, $Lie(G)$ the Lie algebra of right invariant C-derivations on $Q(G)$ and view $Q(G)/C$ as a Δ-field extension with $\Delta=Lie(G)$. Then $Gal_{\Delta}(Q(G)/C)=G(C)$ where $G(C)$ is viewed as acting on $Q(G)$ via right translations. Moreover we have the usual Galois correspondence between intermediate Δ-fields and algebraic subgroups of G.

(3.6) Recall the following definition (cf. [Kol_1]). We say that a Δ-field extension F/K is a Picard-Vessiot extension (respectively an abelian extension) if it is strongly normal and $G_{F/K}$ is a linear algebraic group (respectively an abelian variety). We shall avoid the term "abelian" in denoting commutativity of groups.

(3.7) Galois theory of Picard-Vessiot extension has been generalized by Pommaret [Pomm]. It would be interesting to dispose of theorems similar to our results in (III.2) (III.3) for Pommaret's theory instead of Kolchin's Picard-Vessiot theory.

B. Constrained extensions.

Constrained extensions will appear in our setting only for technical reasons (e.g in the proof of (III.2.6)). We give below a rough sketch of their theory which is due to Kolchin [Kol_4].

(3.9) Let F/K be a partial Δ-field extension and $a=(a_1,\dots,a_k) \in F^k$. One says that a is constrained over K if there exists B

$\in K\{y_1,\dots,y_k\}$(=ring of differential polynomials) with $B(a)\neq 0$ and $B(a')=0$ for all non-generic Δ-specialisations a' of a over K (i.e. for all a' such that the ideal $\{F\in K\{y_1,\dots,y_k\};\ F(a)=0\}$ is strictly contained in the ideal $\{F\in K\{y_1,\dots,y_k\};\ F(a')=0\}$).
F/K is called a constrained extension if every finite family of elements in F is constrained over K. If F is generated as a Δ-field extension of K by $a_1,\dots,a_k$ then F/K is constrained if and only if $(a_1,\dots,a_k)$ is constrained over K.

One can prove that for any constrained extension F/K the extension F^{Δ}/K^{Δ} is algebraic.

A partial Δ-field is called constrainedly closed if it has no non-trivial constrained extension. If K is constrainedly closed then the following holds: if p is a Δ-ideal in $A=K\{y_1,\dots,y_k\}$ and if $B\in A\setminus p$ then there exists $a\in K^k$ with $B(a)\neq 0$ and $F(a)=0$ for all $F\in p$. This immediately implies the following "theorem of zeros" which will be needed later:

(3.10) PROPOSITION. Let F/K be a partial Δ-field extension with F constrainedly closed and let $V\longrightarrow \operatorname{Spec} K$ be a morphism of partial Δ-schemes such that V is a scheme of finite type over K. Then there exists a commutative diagram of Δ-schemes:

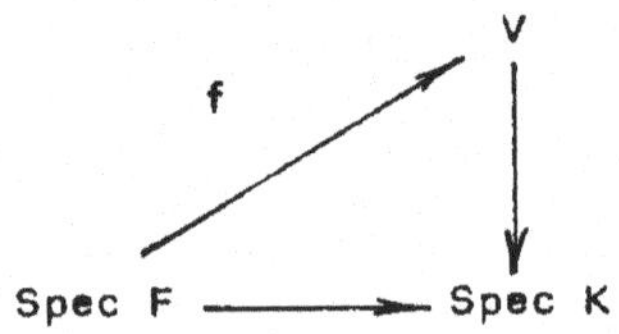

Proof. We may suppose $V=\operatorname{Spec} R$, $R=A/I$, $A=K\{y_1,\dots,y_k\}$, I being a Δ-ideal in A. Then put $A'=F\{y_1,\dots,y_k\}$ and note that IA' is a proper Δ-ideal of A'. By 2) in (2.1) applied to A'/IA' there is a prime Δ-ideal p in A' containing IA' hence by (3.9) there exists $a\in F^k$ annihilating every member of p. This a yelds the morphism f in the diagram above.

(3.11) We will also need the following result from [Kol$_4$]: if K is a partial Δ-field then there exists a constrained extension F/K with F constrainedly closed. In particular any partial Δ-field K with K^{Δ} algebraically closed admits a Δ-field extension F/K such that F is constrainedly closed and $F^{\Delta}=K^{\Delta}$.

In fact much more is true: one can define "constrained closures" of partial Δ-fields and prove uniqueness results for them but we won't need these facts. The following fact won't be used either but worths being noted here:

(3.12) PROPOSITION [Kol$_4$]. Any strongly normal extension is a constrained extension.

C. Logarithmic derivatives and G-primitives.

One of the main features in Kolchin's theory is that strongly normal extensions can be described "modulo Galois cohomology" by means of "Δ-fields arrising from algebraic groups" more precisely by what Kolchin calls G-primitive extensions. A hint about this situation is given by (3.3). In their turn G-primitive extensions are defined by means of "logarithmic derivative". One of the ideas in (II.3) will be to use the logarithmic derivative to solve descent problems. Since in (II.3) we are interested (for reasons which will become apparent there) in general rather than in partial Δ-fields it is convenient to consider here logarithmic derivatives and G-primitives for the "general" rather than "partial" case.

(3.13) First some notational conventions. Let $C\subset K\subset F$ be field extensions and X a variety over C. We shall always identify the set of K-points X(K) of X with a subset of X(F). For any $\alpha\in X(F)$ $\alpha:\operatorname{Spec} F\longrightarrow X$ there is a natural morphism (still denoted by α) $\operatorname{Spec} F\longrightarrow X\otimes_C K$ and we put

$\mathcal{O}_{X,\alpha}$ = local ring of X at $\mathrm{Im}(\alpha)\in X$

$\mathcal{O}_{X\otimes_C K,}$ = local ring of $X\otimes_C K$ at $\mathrm{Im}(\alpha)\in X\otimes_C K$

$\alpha^*: \mathcal{O}_{X\otimes K,\alpha} \longrightarrow F$ the induced ring homomorphism

$K(\alpha) = \mathrm{Im}(\alpha^*)$

(3.14) Now in (3.13) let X be a connected algebraic group over C, with C algebraically closed and write G instead of X. Take $x,y\in G(K)$ and define as usual

$$L_x, R_y: G\otimes K \longrightarrow G\otimes K, \quad L_x(y)=xy, \quad R_y(x)=xy$$

$$L_x^*, R_y^*: Q(G\otimes K) \longrightarrow Q(G\otimes K), \ L_x^* u = uL_x, \quad R_y^* u = uR_y$$

$$L_x^{**}, R_y^{**}: \mathrm{Der}(Q(G\otimes K)) \longrightarrow \mathrm{Der}(Q(G\otimes K))$$

$$L_x^{**}D = (L_x^*)^{-1} D L_x^*$$

$$R_y^{**}D = (R_y^*)^{-1} D R_y^*$$

Recall that we defined

$$\mathrm{Lie}(G) = \mathrm{Lie}_C(G) = \{D\in \mathrm{Der}_C(Q(G));\ R_y^{**}D = D \ \text{ for all } \ y\in G(C)\}$$

Note that $\mathrm{Lie}_C(G) \subset \mathrm{Der}_C(\mathcal{O}_G)$ and equality holds for abelian varieties. Put $\mathrm{Lie}_K(G) = \mathrm{Lie}_C(G)\otimes_C K$; it is a Lie K-algebra.

(3.15) Suppose now in (3.14) that K is a Δ-field and $C\subset K^\Delta$. Then for each $\delta\in\Delta$ one defines a map (called logarithmic derivative cf. [Kol_1] p.394)

$$G(K) \xrightarrow{\ \ell\delta\ } \mathrm{Lie}_K(G)$$

as follows: for any $\alpha\in G(K)$ we let $\ell\delta\alpha$ be the unique element in $\mathrm{Lie}_K(G)$ for which the two maps

$$\mathcal{O}_{G\otimes K,\alpha} \xrightarrow{\ \ell\delta\alpha\ } \mathcal{O}_{G\otimes K,\alpha} \xrightarrow{\ \alpha^*\ } K \quad \text{and}$$

$$\mathcal{O}_{G\otimes K,\alpha} \xrightarrow{\ \alpha^*\ } K \xrightarrow{\ \delta_K\ } K$$

agree on $\mathcal{O}_{G,\alpha}$ (this makes sense because the inclusion $\mathcal{O}_{G,\alpha}\subset\mathcal{O}_{G\otimes K}$, followed by $\delta_K\alpha^*$ is a C-derivation of $\mathcal{O}_{G,\alpha}$ into K which uniquely extends to a K-derivation of $\mathcal{O}_{G\otimes K,\alpha}$ into K; but such a derivation is of the form α^*D for a unique $D\in \mathrm{Lie}_K(G)$). By the very definition remark that if F/K is a Δ-field extension then we have a commutative diagram

$$\begin{array}{ccc} G(K) & \xrightarrow{\ \ell\delta\ } & \mathrm{Lie}_K(G) \\ \downarrow & & \downarrow \\ G(F) & \xrightarrow{\ \ell\delta\ } & \mathrm{Lie}_F(G) \end{array}$$

(3.16) LEMMA [Kol_1 p.353]. If $\alpha\in G(K)$ then $\ell\delta\alpha = L_\alpha^{**}\delta_K^{*}-\delta_K^{*}$ where δ_K^{*} denotes as usual the trivial lifting of δ_K to $Q(G\otimes K)=Q(Q(G)\otimes K)$.

Proof. Remark that $D:=L_\alpha^{**}\delta_K^{*}-\delta_K^{*}$ belongs to $\mathrm{Lie}_K(G)$; indeed it vanishes on $1\otimes K$ and $R_\beta^{**}D=D$ for all $\beta\in G(\mathbb{C})$. Now one just has to look at the following commutative diagram:

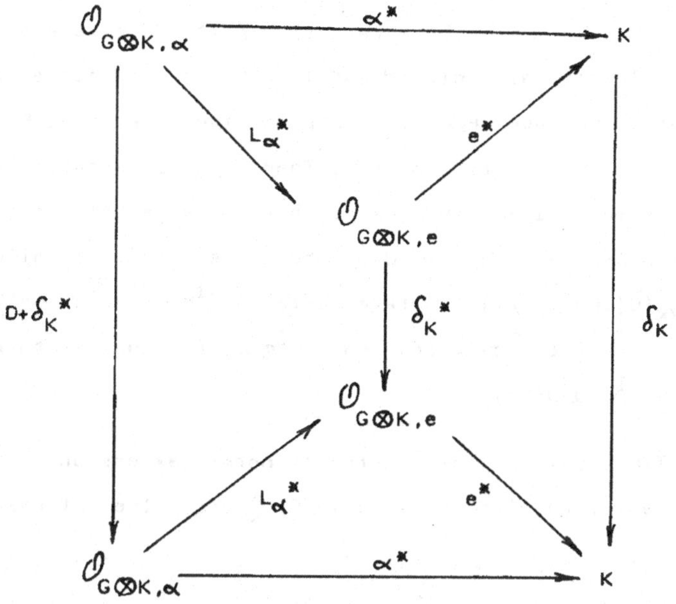

where $e \in G(K)$ is the identity.

(3.17) Let F/K be a Δ-field extension, $C \subset K^{\Delta}$ an algebraically closed field and G a connected algebraic group. Take $\alpha \in G(F)$; we say that α is a G-primitive over K if

$$\ell\delta\alpha \in \mathrm{Lie}_K(G) \quad \text{for all} \quad \delta \in \Delta$$

We say that F/K is a G-primitive extension if $F=K(\alpha)$ for some G-primitive $\alpha \in G(F)$ over K. F/K will be called a full G-primitive extension if in addition $\mathrm{tr.deg.}F/K=\dim G$. By the very definition of logarithmic derivative one sees that if F/K is a full G-primitive extension with G-primitive $\alpha \in G(F)$ then α induces a K-isomorphism $\alpha^*: Q(G \otimes K) \longrightarrow F$ such that

$$(\alpha^*)^{-1}\delta_F\alpha^* = \delta_K^* + \ell\delta\alpha \qquad \text{for all} \quad \delta \in \Delta$$

where δ_K^* is the trivial lifting of δ_K to $Q(G \otimes K)$ and $\ell\delta\alpha$ is viewed as an element of $\mathrm{Der}(Q(G \otimes K))$.

The connection between strong normality and G-primitives is given by the following results due to Kolchin [Kol_1] pp.419 and 426:

(3.18) PROPOSITION. Let F/K be a partial Δ-field extension with $C:=K^{\Delta}=F^{\Delta}$ algebraically closed and let G be a connected algebraic group over C. Suppose F/K is a G-primitive extension, $F=K(\alpha)$ with $\alpha \in G(F)$ a G-primitive over K. Then F/K is strongly normal. Moreover there is an injective homomorphism of algebraic groups $G_{F/K} \longrightarrow G$ which at the level of C-points is described as follows: an element $\sigma \in G_{F/K}(C)=\mathrm{Gal}_{\Delta}(F/K)$ is taken into $\alpha^{-1}\sigma\alpha$ (here $\mathrm{Gal}_{\Delta}(F/K)$ naturally acts on $G(F)$ so $\sigma\alpha \in G(F)$ so $\alpha^{-1}\sigma\alpha \in G(F)$ but it turns out that in fact $\alpha^{-1}\sigma\alpha \in G(C)$).

(3.19) PROPOSITION. Let F/K be a strongly normal extension with K algebraically closed. Then F/K is a full $G_{F/K}$-primitive extension.

(3.20) COROLLARY. Let F/K be a Δ-function field which is strongly normal. Then there exists a finite extension K_1/K such that K_1F/K_1 is a full $G_{F/K}$-primitive extension. Furthermore $G_{F/K}=G_{K_1F/K_1}$.

(3.21) Using the concept of V-primitive instead of that of G-primitive (where V is a principal homogeneous space for G) Kolchin proved in [Kol_1] p.430 a result which is more precise than (3.18)+(3.19) above; we shall not need this result in what follows.

CHAPTER II. DIFFERENTIAL DESCENT THEORY.

The primary goal of the theory we are going to develop in this chapter is to obtain higher dimensional analogs of the main results in [Mtd]. But our theory has also a different flavour, namely it should be viewed as an infinitesimal analog of the theory of fields of moduli in the sense of Shimura [Sh_1] and Matsusaka [Mtk] cf. [Koiz]; automorphisms of the universal field in their theory are replaced here by derivations.

An earlier (much weaker) version of the present theory is contained in [Bu_2]. Our main ingredients will be: deformation theory (of polarized algebraic varieties and compact analytic spaces)cf.(II.1) and Kolchin's logarithmic derivative cf. (II.2).

1. Descent of projective Δ-varieties.

If V is a variety over K and C is a subfield of K one says that C is a field of definition for V (or that V is defined over C or that it descends to C or that it is rational over C) if there exist a variety X over C and a K-isomorphism $\sigma:V\longrightarrow X\otimes_C K$ (which will be called a descent isomorphism). Our first main result is the following:

(1.1)THEOREM.Let K be an algebraically closed Δ-field and V a projective Δ-variety over K. Then K^Δ is a field of definition for V.

This theorem was proved in [Bu_2] under the assumption that V is non-singular. The proof there may be adapted to work in the singular case too as we shall explain below. We shall reproduce however many arguments from [Bu$_2$] for the convenience of the reader. To get a better understanding of (1.1) note for instance that if Δ has one element only and if $\mathrm{tr.deg.}K/K^{\Delta}=1$ then it may be shown that $V \longrightarrow \mathrm{Spec}\, K$ is the geometric general fibre of a projective morphism $f: X \longrightarrow S$ where X, S are varieties over K^{Δ}, $\dim(S)=1$ and δ_K, δ_V "come" from vector fields on S and X respectively; our theorem reduces then to Ehresmann's theorem on foliations which are transverse to a proper map [J$_1$] p.210 (in fact one needs for this $K^{\Delta}=\mathbb{C}$ but one can reduce the problem to this situation). The case $\mathrm{card}\,\Delta = \mathrm{tr.deg.}K/K^{\Delta} = 1$ was also considered in [MD$_1$], [MD$_2$] as a preliminary for descent problems related to "Mordell conjecture over function fields". No such reduction to Ehresmann's theorem is possible in the general case cf. (IV.2), not even in the case $\mathrm{card}\,\Delta=1$ and $2 \leq \mathrm{tr.deg.}K/K^{\Delta} < \infty$. As we shall explain in (IV.2) one is forced on the other hand to consider the general case when dealing with algebraic differential equations with meromorphic coefficients. Note also that Theorem (1.1) was implicitely proved in [Mtd] in the case $\mathrm{card}\,\Delta=1$, $\dim(V)=1$ and arbitrary $\mathrm{tr.deg.}K/K^{\Delta}$, but the proof there extensively used the hypothesis $\dim(V)=1$ (it used Weierstrass normal forms of elliptic curves and Weierstrass points on curves of genus ≥ 2) so cannot be extended to the case $\dim(V) \geq 2$.

(1.2) The way in which (1.1) will apply to Galois theory of partial Δ-fields will become clear in (III.2). There is however another application of (1.1) we would like to discuss here. Given a variety V over a field K there is a natural set of derivation operators $\Delta(V)$ making K a $\Delta(V)$-field and V a $\Delta(V)$-variety over K namely

$$\Delta(V) = \{\delta \in \mathrm{Der}(\mathcal{O}_V) ;\ \delta(K) \subset K\}$$

Roughly speaking it is the richest structure of Δ-variety one can introduce on V. Certainly V is not a partial $\Delta(V)$-variety. Note that $\Delta(V)$ has a natural structure of Lie $\mathbb{Q}$-algebra and of K-vector space. Note also that if V is projective then by (I.2.5) the condition $\delta(K)\subset K$ from the definition of $\Delta(V)$ is superfluous hence $\Delta(V)$ simply equals $Der(\mathcal{O}_V)$. We have the following:

(1.3) COROLLARY. Let K be an algebraically closed field and let V be a projective variety over K. Then the set of all algebraically closed fields of definition for V has a smallest element which equals $K^{\Delta(V)}$. Moreover we have a split exact sequence of Lie $\mathbb{Q}$-algebras and K-vector spaces

$$0\longrightarrow Der_K(\mathcal{O}_V)\longrightarrow Der(\mathcal{O}_V)\longrightarrow Der_{K^{\Delta(V)}}(K)\longrightarrow 0$$

Proof. Let's prove that for any algebraically closed field of definition K_1 for V we have $K^{\Delta(V)}\subset K_1$. It is sufficient to prove that that $Der_{K_1}(K)\subset Im(\Delta(V)\longrightarrow Der(K))$. Now if $\sigma:V\simeq V_1\otimes_{K_1}K$ and $\delta\in Der_{K_1}(K)$ then δ is the restriction of its own trivial lifting δ^* to $V_1\otimes_{K_1}K$ so $\delta\in Im(\Delta(V)\longrightarrow Der(K))$. We immediately conclude by (1.1) that $K^{\Delta(V)}$ is the smallest algebraically closed field of definition for V. The splitting of the exact sequence in the statement of (1.3) is given by the map $Der_{K^{\Delta(V)}}(K)\longrightarrow Der(\mathcal{O}_V)$ which takes δ into $(\sigma^*)^{-1}\delta^*\sigma^*$ where $\sigma^*:Q(V_1\otimes_{K_1}K)\longrightarrow Q(V)$ is induced by σ.

Let's remark that there are examples (due to Shimura) of non-singular projective curves V for which the set of all fields of definition doesn't have a smallest element, see (1.4) below.

(1.4) Let's quickly recall some facts about fields of moduli [Koiz] [Sh_2] and compare them with our Corollary (1.3). For simplicity we shall discuss only the case of curves. Given a non-singular projective curve V over an algebraically closed field K one says that a subfield k of K is a field of moduli for V if for any automorphism $\sigma\in Aut(K)$ we have $V^{\sigma}\simeq V$ if and only if $\sigma\in Aut(K/k)$. Here

$V^{\sigma} \longrightarrow \operatorname{Spec} K$ is simply the variety $V \longrightarrow \operatorname{Spec} K \xrightarrow{\operatorname{Spec}\sigma} \operatorname{Spec} K$ and the isomorphism $\simeq$ is over K. It is known [Koiz] that the field of moduli always exists and it equals the intersection of all fields of definition for V. On the other hand Shimura [Sh_2] gave an example of (hyperelliptic) curve whose field of moduli is not a field of definition: it is the curve V over $K=\mathbb{C}$ with plane equation

$$y^2=a_0x^m+\sum_{r=1}^{m}(a_r^{\sigma}x^{m+r}+(-1)^r a_r x^{m-r})$$

with m odd, $a_m=1$, $a_0\in\mathbb{R}$, a_j^{σ}=complex conjugate of a_j, where we assume that the coefficients $a_0,a_1,\dots,a_{m-1}$ are chosen generic with the above properties. In particular for V above the set of all fields of definition does not have a smallest element.

Now we would like to explain why $K^{\Delta(V)}$ in Corollary (1.3) should be viewed as an infinitesimal analog of the concept of field of moduli. Returning to Shimura's definition we see that if k is the field of moduli for the curve V then $k=K^{G(V)}$ where

$$G(V)=\{\sigma\in \operatorname{Aut}(V/\mathbb{Q})\ ;\ \sigma^*(K)=K\}$$

Here $\operatorname{Aut}(V/\mathbb{Q})$ is the group of all automorphisms of the scheme V (not necessarily over K !); $\sigma^*:\mathcal{O}_V\longrightarrow\sigma_*\mathcal{O}_V$ is the induced morphism of sheaves of rings; $K^{G(V)}$ denotes as usual the field of invariants of the group $\operatorname{Im}(G(V)\longrightarrow\operatorname{Aut}(K))$ in K. So $K^{\Delta(V)}$ appears as an analog of $K^{G(V)}$ (automorphisms being replaced by derivations). What the general theory of fields of moduli [Koiz] does is to compare fields of moduli with fields of definition. We shall do the same but replacing fields of moduli with their infinitesimal analogs and fields of definition with algebraically closed fields of definition.

The proof of Theorem (1.1) will be intimately related to the proof of the following

(1.5) THEOREM. Let $f:X \longrightarrow S$ be a projective surjective morphism of varieties over an algebraically closed field C. Suppose f has geometrically integral fibres. Then there exists a diagram of varieties over C with cartesian squares:

$$\begin{array}{ccccc} X & \longleftarrow & X' & \longrightarrow & X'' \\ \downarrow & & \downarrow & & \downarrow \\ S & \longleftarrow & S' & \longrightarrow & S'' \end{array}$$

where 1) $S' \longrightarrow S$ is a quasi-finite map

2) $S' \longrightarrow S''$ is sujective and S'' is affine and smooth, $S''=\mathrm{Spec}\,A$.

3) $X'' \longrightarrow S''$ is projective and the Kodaira-Spencer map

$$\rho:\mathrm{Der}_C(A) \longrightarrow \mathrm{Ext}^1(\Omega_{X''/S''},\mathcal{O}_{X''})$$

is an injection of A-modules.

(1.6)The result above (for $C=\mathbb{C}$) was proved by Viehweg [Vie] under quite restrictive assumptions: first it was supposed that f is smooth but, what is more important, it was assumed that the composition

$$T_sS \longrightarrow H^1(X_s,\underline{\mathrm{Der}}_C(\mathcal{O}_{X_s})) \longrightarrow \mathrm{Hom}_C(H^{n0},H^{n-1,1})$$

is injective for all sufficiently general $s \in S(C)$ where T_sS=Zariski tangent space of S at s, $X_s=f^{-1}(s)$, $H^{pq}=H^q(X_s,\Omega^p_{X_s/C})$, the first arrow is the usual Kodaira-Spencer map and the second map is the "infinitesimal Torelli map" given by cup-product ($n=\dim(X_s)$). Viehweg's proof was based on the "algebraicity" of the period map. Our proof will be quite different.

The interplay between proofs of Theorems (1.1) and (1.5) is a little bit tricky. We shall prove them is four steps as follows:

Step 1. Theorem (1.5) for $C=\mathbb{C}$.

Step 2. Theorem (1.1) for $\mathbb{C}\subset K^{\Delta}$.

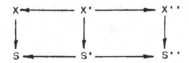

Step 3. Theorem (1.1) for arbitrary K^{Δ}.

Step 4. Theorem (1.5) for arbitrary C.

Note that in Corollary (1.3) we always have $\mathrm{tr.deg.}K^{\Delta(V)}/\mathbb{Q}<\infty$ just because V has always a field of definition finitely generated over $\mathbb{Q}$. On the other hand the proof of our theorems will be done first over $\mathbb{C}$ by invoking an analytic argument !

(1.6) A remark: the projectivity hypothesis in (1.1) and (1.3) cannot be removed as shown by an easy example in [Bu_3]. However we shall be able to prove a generalisation of (1.1) and (1.3) for the case of "open varieties" by assuming a "logarithmic behaviour of derivations at infinity" , see (3.6) below. But this will require an additional idea namely the use of the logarithmic derivative.

(1.7) Another remark: Theorem (1.5) is easier for those $f:X\longrightarrow S$ whose fibres have a"good theory of moduli" (in the sense for instance that there exists an algebraic space which is a coarse moduli space for the polarized fibres). A fact is that our applications to differential Galois theory in (III.2),(III.3) will force us to apply (1.1) and hence indirectly (1.5) in cases where coarse moduli spaces hardly exist (for instance for varieties which turn out aposteriori to be completions of homogenous spaces of linear algebraic groups,hence unirational).

(1.8) It worths giving the following consequences of (1.3). Let K be an algebraically closed field and C an algebraically closed subfield of K. Then the following hold:

a) If V and W are normal projective varieties over K which are isomorphic in codimension 1 (i.e. there exist open subsets $V_{o}\subset V$, $W_{o}\subset W$ with $V_{o}\simeq W_{o}$, $\mathrm{codim}_{V}(V\setminus V_{o})\geq 2$, $\mathrm{codim}_{W}(W\setminus W_{o})\geq 2$) and if V descends to C then so does W.

b) If $f:V\longrightarrow W$ is a morphism of projective varieties over K such that $f_{*}\mathcal{O}_{V}=\mathcal{O}_{W}$ and V descends to C, then so does W.

To prove a) note that $\Delta(V)$ identifies with $\Delta(W)$ by (I.2.3) and we conclude by (1.3). To prove b) note that exactly as in (I.2.5) we have a natural restrictionm map $Der(\mathcal{O}_V) \longrightarrow Der(\mathcal{O}_W)$ and we are done again by (1.3).

Note however that both a) and b) above can be proved using a much weaker form of (1.1) (they are implied by an easy consequence of Ehresmann's theorem).

Let's start proving (1.1) and (1.5). We begin by an easy preparation on isomorphisms of polarized varieties.

(1.9) Let $X \longrightarrow Z$, $Y \longrightarrow Z$ be projective morphisms of noetherian schemes and consider the functor

$$\underline{Isom}_{X/Z,Y/Z} : \left\{ \begin{array}{l} \text{locally noetherian} \\ Z\text{-schemes} \end{array} \right\} \longrightarrow \{\text{sets}\}$$

$$\underline{Isom}_{X/Z,Y/Z}(T) = Isom_T(X \times_Z T, Y \times_Z T)$$

By [FGA] this functor is representable by an open subscheme of the Hilbert scheme $Hilb_{X \times_Z Y/Z}$.

Recall also that if X is a projective variety over an algebraically closed field C then the Neron-Severi group $NS(X) = Pic(X)/Pic^{\tau}(X)$ is at most countable (for the definition of Pic^{τ} see [FGA]; countability of $NS(X)$ is an immediate consequence of representability theorems in [FGA]). An element of $NS(X)$ having an ample representative in $Pic(X)$ will be called in what follows a polarisation on X. If Y is another projective variety over C, $L \in Pic(X)$, $M \in Pic(Y)$ we write $(X,L) \overset{\tau}{\sim} (Y,M)$ iff there exists a C-isomorphism $f: X \longrightarrow Y$ such that $L^{-1} \otimes f^* M \in Pic^{\tau}(X)$.

(1.10) LEMMA. Let $f: Y \longrightarrow T$ be a flat surjective projective morphism of schemes with geometrically integral fibres, T locally noetherian. Let $L \in Pic(Y)$. Then there exists an open subscheme T' of T

such that if F is any algebraically closed field we have

$$T'(F)=\{t\in T(F);\ L_t\in \mathrm{Pic}^{\tau}(Y_t)\}$$

(where if $t:\mathrm{Spec}\,F\longrightarrow T$ then of course $Y_t=Y\times_T \mathrm{Spec}\,F$ and $L_t=$ inverse image of L on Y_t).

Proof. First we may reduce ourselves to the case when f has a section. Then one applies the results in [FGA] or [Mum] p.23 on representability of the relative Picard functor.

(1.11) LEMMA. Let $X^1\longrightarrow Z$, $X^2\longrightarrow Z$ be flat surjective projective morphisms of noetherian schemes having geometrically integral fibres. Let $L^1\in \mathrm{Pic}(X^1)$, $L^2\in \mathrm{Pic}(X^2)$ ample relative to Z. Then there is a constructible set $Z'\subset Z$ such that if F is any algebraically closed field then

$$Z'(F)=\{z\in Z(F);\ (X^1{}_z,L^1{}_z)\overset{\tau}{\sim}(X^2{}_z,L^2{}_z)\}$$

where $Z'(F)$ is by definition the set of all $z:\mathrm{Spec}\,F\longrightarrow Z$ such that $z(\mathrm{Spec}\,F)\in Z'$.

Proof. Let T be the object representing $\underline{\mathrm{Isom}}_{X^1/Z,X^2/Z}$ consider the cartesian diagrams

$$\begin{array}{ccc} Y^i & \xrightarrow{u_i} & X^i \\ \downarrow & & \downarrow \\ T & \xrightarrow{p} & Z \end{array}$$

$i=1,2$ and let $\varphi:Y^1\longrightarrow Y^2$ be the universal T-isomorphism. Let $T'\subset T$ be the open set defined in (1.10) and corresponding to $L=M^1\otimes(\varphi^*(M^2))^{-1}$ where $M^i=u_i{}^*(L^i)$, $i=1,2$. Finally put $Z'=p(T')$. It is easy to check that T' is contained in finitely many connected components of the Hilbert scheme of $X^1\times_Z X^2/Z$ (just compute Hilbert polynomials cf. [Bu_2] for instance) hence Z' is constructible. Then one easily checks that Z' enjoys the property in the statement.

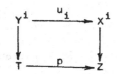

(1.12) LEMMA. Let $X \longrightarrow S$ be a flat surjective projective morphism of noetherian schemes having geometrically integral fibres. Let $L \in \mathrm{Pic}(X)$ ample relative to S. Then there is a constructible subset R of $S \times S$ such that for any algebraically closed field F and any $s_1, s_2 \in S(F)$ we have

$$(s_1, s_2) \in R(F) \quad \text{iff} \quad (X_{s_1}, L_{s_1}) \overset{\tau}{\sim} (X_{s_2}, L_{s_2})$$

Proof. Put $Z = S \times S$, consider the cartesian diagrams

$$\begin{array}{ccc} X^i & \xrightarrow{\ u_i\ } & X \\ \downarrow & & \downarrow \\ Z & \xrightarrow{\ p_i\ } & S \end{array}$$

$i=1,2$ with p_i=the i-th projection and put $L^i = u_i^* L$. Then aplly (1.11).

Now we shall need the following result on the existence of "birational quotient" of varieties by a constructible equivalence relation:

(1.13) LEMMA. Let S be a variety over an algebraically closed field C and $R \subset S \times S$ a constructible set such that $R(C)$ is an equivalence relation on $S(C)$. Then there exists a Zariski open set $S_o \subset S$ and a morphism $\psi : S_o \longrightarrow M$ of C-varieties such that for any $s \in S_o(C)$ we have

$$\psi^{-1}\psi(s) = \left\{ t \in S_o(C);\ (s,t) \in R(C) \right\}$$

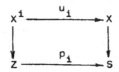

Remark. It seems that the general yoga of birational quotients related to groupoids from [SGA 3] does not apply (at least directly) to this context since there seems to be no groupoid behind our equivalence relation in general (R may not be a subscheme of $S \times S$ apriori; remark that if R and S are as in (1.12) and if $X \longrightarrow S$ in (1.12) is a smooth morphism whose fibres are non-ruled varieties then by a theorem of Matsusaka [Mtk] R is closed in $S \times S$; however in our main applications in (III.2),(III.3) the fibres of $X \longrightarrow S$ will often be ruled). An ad-hoc argument inspired from [Re]p.406 may be given as follows:

Proof ([Bu_2]). Embed S as a locally closed subscheme is some projective space P, write $R=R_1\cup\ldots\cup R_N$ with R_i irreducible and locally closed in $S\times S$, let $\overline{R}$ and $\overline{R}_i$ be the closures of R and R_i in $P\times S$ and suppose $\overline{R}=\overline{R}_1\cup\ldots\cup\overline{R}_n$, $n\leq N$ such that there is no inclusion $\overline{R}_i\subset\overline{R}_j$ for $1\leq i<j\leq n$. Let $f:\overline{R}\longrightarrow S$ and $f_i:\overline{R}_i\longrightarrow S$ be the natural projections; there is open set $S_o\subset S$ such that f_i are flat above S_o. Shrinking S_o we may suppose that:

1) $f_i^{-1}(y)$ is reduced at the generic points of its components,

2) $f_i^{-1}(y)\cap R_i$ is open dense in $f_i^{-1}(y)$,

3) No component of $f_i^{-1}(y)$ is contained in a component of $f_j^{-1}(y)$ provided $y\in S_o(C)$ and $1\leq i<j\leq n$.

Let $\overline{Z}_i(y)$ be the projection of $f_i^{-1}(y)$ into P and $Z_i(y)$ be the projection of $f_i^{-1}(y)\cap R_i$ into P. Then $y\longmapsto\overline{Z}_i(y)$ induces a morphism $S_o\longrightarrow Hilb_P$ and hence by [Ang] a morphism

$$y\longmapsto \text{cycle}(\overline{Z}_i(y))$$

$$S_o\longrightarrow \text{Chow}(n_i,d_i)$$

where $\text{Chow}(n_i,d_i)$ is the Chow variety of cycles in P of degree d_i and dimension n_i. Consider the map

$$\Psi:S_o\longrightarrow\prod_{i=1}^{n}\text{Chow}(n_i,d_i)\longrightarrow M=\prod_{q=1}^{n^*}\text{Chow}(q,\sum_{n_i=q}d_i)$$

with $n^*=\max(n_i)$. This Ψ is good for our purpose. The point is that if $\Psi(s)=\Psi(t)$ for $s,t\in S_o(C)$ then for all q we have

$$\begin{array}{ccc}\cup Z_i(s) & & \cup Z_i(t)\\ \downarrow & & \downarrow\\ \cup\overline{Z}_i(s) & = & \cup\overline{Z}_i(t)\end{array}$$

where the vertical arrows are open dense immersions (by condition 2) above) and i is running through all indices for which $n_i=q$. Consequently $(\cup Z_i(s))\cap(\cup Z_i(t))\neq\phi$ hence there exists $y\in P(C)$ and indices $1\leq i\leq j\leq n$ such that $(y,s)\in R_i(C)$ and $(y,t)\in R_j(C)$. Since $R(C)$ is an equivalence relation we get $(s,t)\in R(C)$. Conversely if $(s,t)\in(R\cap(S_o\times S_o))(C)$ we get

$$\bigcup_{i=1}^{N} Z_i(s) = \bigcup_{i=1}^{N} Z_i(t)$$

But for any $k\geq n+1$ and $y\in S_o(C)$, $Z_k(y)\subset\overline{Z}_i(y)$ for some $i\leq n$. We get

$$\bigcup_{i=1}^{n} \overline{Z}_i(s) = \bigcup_{i=1}^{n} \overline{Z}_i(t)$$

By conditions 1) and 3) above we get $\psi(s)=\psi(t)$ and the lemma is proved.

(1.14) We shall need some general facts about the relative Ext [BPS] which we now briefly recall. Let $f:X\longrightarrow S$ be a proper flat morphism of noetherian schemes, $\mathcal{F}$ and $\mathcal{G}$ coherent sheaves on X which are flat over S and suppose $S=\mathrm{Spec}\ A$. One defines $\underline{\mathrm{Ext}}^i(f,\mathcal{F},\mathcal{G})$ to be the sheaf associated to the presheaf

$$U\longmapsto \mathrm{Ext}^i_{f^{-1}(U)}(\mathcal{F}|_{f^{-1}(U)},\mathcal{G}|_{f^{-1}(U)})$$

It is a sheaf of A-modules and it is easy to check that it equals $(\mathrm{Ext}^i(\mathcal{F},\mathcal{G}))^{\sim}$. Using this remark and the results in [BPS] we get that the following properties hold (for the second we suppose that A is regular):

1) $y\longmapsto \dim_{k(y)}(\mathrm{Ext}^i(\mathcal{F}_y,\mathcal{G}_y))$ is upper semicontinous,

2) If the function above is constant then $\mathrm{Ext}^i(\mathcal{F},\mathcal{G})$ is locally free and for all $y\in S$ we have natural isomorphisms

$$\mathrm{Ext}^i(\mathcal{F},\mathcal{G})\otimes k(y)\simeq \mathrm{Ext}^i(\mathcal{F}_y,\mathcal{G}_y)$$

where $k(y)$ denotes of course the residue field of S at y.

3) If $u:T\longrightarrow S$ is a flat morphism then

$$u^*(\underline{\mathrm{Ext}}^i(f,\mathcal{F},\mathcal{G}))\simeq\underline{\mathrm{Ext}}^i(g,u^*\mathcal{F},u^*\mathcal{G})$$

where $g: X \times_S T \longrightarrow T$ is the induced morphism.

(1.15) We shall also need an easy remark concerning deformations of compact complex spaces (for background see [Dou], [Bing]). Recall first that if X_0 is a reduced compact complex space then by [Sch] the space of infinitesimal deformations of X_0 identifies with $\mathrm{Ext}^1(\Omega_{X_0}, \mathcal{O}_{X_0})$ (if $X_0 = V^{an}$ with V a projective variety over $\mathbb{C}$ then by GAGA the Ext above identifies with the algebraic one $\mathrm{Ext}^1(\Omega_{V/\mathbb{C}}, \mathcal{O}_V)$). Now the remark we need is the following. Suppose $f: X \longrightarrow S$ is a proper flat morphism of complex spaces having reduced fibres. Assume also that S is smooth and:

1) $s \longmapsto \dim \mathrm{Ext}^1(\Omega_{X_s}, \mathcal{O}_{X_s})$ is constant for all $s \in S$,

2) The Kodaira-Spencer map $\rho_s : T_s S \longrightarrow \mathrm{Ext}^1(\Omega_{X_s}, \mathcal{O}_{X_s})$ is zero for all $s \in S$.

Then we claim that $f: X \longrightarrow S$ is locally analytically trivial.

The quickest argument for this is via versal deformation and will be given below; note however that one can give a proof which avoids the existence of versal deformation (in the spirit of [Ker]). Take $s_0 \in S$ and let $Y \longrightarrow (B, b_0)$ be a versal deformation for X_{s_0}. There exists a map $\varphi: (S, s_0) \longrightarrow (B, b_0)$ inducing the family $X \longrightarrow (S, s_0)$ from $Y \longrightarrow (B, b_0)$. It is sufficient to prove that $\mathcal{O}_{B,b_0} \longrightarrow \mathcal{O}_{S,s_0}$ factors through $\mathcal{O}_{B,b_0} \longrightarrow \mathbb{C}$. Since S is smooth it is sufficient to check that the tangent map $T_s\varphi = 0$ for all $s \in S$ close to s_0. We have a commutative diagram

$$\begin{array}{ccc} T_s S & \xrightarrow{\quad \rho_s \quad} & \mathrm{Ext}^1(\Omega_{X_s}, \mathcal{O}_{X_s}) \\ {\scriptstyle T_s\varphi}\downarrow & & \downarrow\wr \\ T_b B & \xrightarrow{\quad \rho_b \quad} & \mathrm{Ext}^1(\Omega_{Y_b}, \mathcal{O}_{Y_b}) \end{array}$$

where $b = \varphi(s)$. Since $Y \longrightarrow (B, b)$ is versal for b close to b_0 [Bing] ρ_b is surjective for b close to b_0. Since

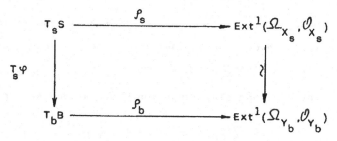

$$\dim_{\mathbb{C}} T_b B \leqslant \dim_{\mathbb{C}} T_{b_o} B$$

for b close to b_o and since ρ_{b_o} is bijective it follows from hypothesis 1) that ρ_b is bijective for all $b \in \varphi(S)$. By hypothesis 2) we get $T_s\varphi = 0$ and we are done.

(1.16) With the preparation above Theorems(1.1) and (1.5) may be proved exactly as in [Bu$_2$]. We shall briefly recall the arguments. First one proves (1.5) in the case $C = \mathbb{C}$. Shrinking S in (1.5) we may suppose f is flat. Fix any $L \in \mathrm{Pic}(X)$ ample relative to S and let $R \subset S \times S$ be the constructible set defined by Lemma (1.12). Then let $\psi : S_o \longrightarrow M$ be the map defined by Lemma (1.13); we may suppose that $S_o = S$ and $\overline{\psi(S)} = M$. Let N be a locally closed subscheme of S étale over M and put $\eta : S_1 = S \times_M N \longrightarrow N$. Clearly η has a section and for any $t \in N(C)$ the set of all $s \in N(C)$ for which X_s is $\mathbb{C}$-isomorphic to X_t (as a non-polarized variety !) is at most countable (here X_s is the fibre of $X \longrightarrow S$ above s). Now exactly as in [Vie] p.576 we may suppose there is a projective morphism $g : Y \longrightarrow N$ (N smooth affine) such that $X_1 := X\ {}_S S_1$ is S_1-isomorphic to $Y \times_N S_1$. This is done by considering a section $N \xrightarrow{\sigma} T \subset S_1$ and making base changes

$$\begin{array}{ccc} X_T & \longrightarrow & X_1 \\ \downarrow & & \downarrow \\ T & \longrightarrow & S_1 \end{array} \qquad \begin{array}{ccc} X_N & \longrightarrow & X_T \\ \downarrow & & \downarrow \\ N & \xrightarrow{\sigma} & T \end{array} \qquad \begin{array}{ccc} X_1^{*} & \longrightarrow & X_N \\ \downarrow & & \downarrow \\ S_1 & \longrightarrow & N \end{array}$$

Then $X_1 \longrightarrow S_1$ and $X_1^{*} \longrightarrow S_1$ are "fiberwise isomorphic" hence they become isomorphic after a quasi-finite base change $S_2 \longrightarrow S_1$ (use representability of $\underline{\mathrm{Isom}}_{X_1/S_1, X_1^{*}/S_1}$ and a Baire-type argument). We replace S_1 by S_2 and claim that the Kodaira-Spencer map

$$\rho : \mathrm{Der}_C(\mathcal{O}_N) \longrightarrow \mathrm{Ext}^1(\Omega_{Y/N}, \mathcal{O}_Y)$$

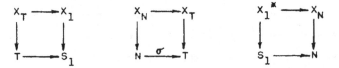

is injective after some shrinking of N. Indeed by (1.14) we may shrink N such that the upper semicontinous function from 1) in (1.14) is constant hence by 2) in (1.14) base change theorems will hold for Ext^1 on N. Suppose ρ was not injective. Shrinking N we may assume that $Ker(\rho)$ contains a line bundle L. By Frobenius L is integrable so there exists a piece of analytic curve Z in N (non-algebraic in general !) whose analytic tangent bundle is L. By (1.15) the analytic family $Y \times_N Z \longrightarrow Z$ is locally trivial in particular all fibres of $Y \longrightarrow N$ above points of Z are $\mathbb{C}$-isomorphic. Since Z is uncountable this leads to a contradiction.

(1.17) Next step is to prove Theorem (1.1) in the case $\mathbb{C} \subset K^{\Delta}$. Using $[J_2]$ p.36 we may write $V \simeq X \times_S \mathrm{Spec}\, K$ where $f: X \longrightarrow S$ is a projective surjective map of varieties over $\mathbb{C}$ having geometrically integral fibres and $\mathrm{Spec}\, K \longrightarrow S$ is a dominant morphism of schemes. Now apply (1.5) and put $K_0 = Q(S'')$ and $V_0 = X'' \times_{S''} \mathrm{Spec}\, K_0$. Since $Q(S')$ is finite over $Q(S)$, $Q(S')$ embeds into K over $Q(S)$ and one can check that V is K-isomorphic to $V_0 \otimes_{K_0} K$ so the only thing we have to prove is that K_0 is contained in K^{Δ}. Now look at the commutative diagram (where $C = \mathbb{C}$) with exact rows and colomns :

$$\begin{array}{ccccc}
 & & 0 & & \\
 & & \downarrow & & \\
 & & Der_{K_0}(K) & & 0 \\
 & & \psi \downarrow & & \downarrow \\
Der_C(\mathcal{O}_V) & \xrightarrow{\varphi} & Der_C(K) & \longrightarrow & Ext^1(\Omega_{V/K}, \mathcal{O}_V) \\
 & & \downarrow & & \wr\downarrow \\
 & & Der_C(K_0) \otimes_{K_0} K & \xrightarrow{\alpha} & Ext^1(\Omega_{V_0/K_0}, \mathcal{O}_{V_0}) \otimes_{K_0} K \\
 & & \downarrow & & \downarrow \\
 & & 0 & & 0
\end{array}$$

Since α is injective $Im(\varphi) = Im(\psi)$ and we are done.

(1.18) Next step is to prove Theorem (1.1) for general K^{Δ}. We may suppose $K^{\Delta} \subset \mathbb{C}$. Look at $\mathbb{C}$ as a Δ-field with zero derivations and put $F=Q(K \otimes_{K^{\Delta}} \mathbb{C})$, $V_F = V \otimes_K F$ in the category of Δ-fields and Δ-varieties respectively. Since $\mathbb{C} = 1 \otimes \mathbb{C} \subset F^{\Delta}$ we get by (1.17) that V_F descends to F^{Δ}. On the other hand V_F descends by its very definition to K. Since by (I.1.2) F^{Δ} and K are linearly disjoint over K^{Δ} it follows by Lemma (1.19) below that V_F descends to K^{Δ} ($V_F \simeq X \otimes_{K^{\Delta}} F$). Representability of $\underline{\mathrm{Isom}}_{V/K, X \otimes_{K^{\Delta}} K/K}$ yelds that V itself descends to K^{Δ}.

(1.19) LEMMA. Let V be a projective variety over an algebraically closed field K and let K_1, K_2 be algebraically closed subfields of K which are linearly disjoint over their intersection $K_o = K_1 \cap K_2$. If V descends to both K_1 and K_2 then it also descends to K_o.

Remark. Aposteriori the statement above will be proved to hold without the linear disjointness assumption, see Corollary (1.3).

Proof ([Bu_2]). Choose an ample $L \in \mathrm{Pic}(V)$. It is possible to find cartesian diagrams

$$\begin{array}{ccc} V & \xrightarrow{\;u_i\;} & X_i \\ \downarrow & & \downarrow f_i \\ \mathrm{Spec}\, K & \longrightarrow & S_i \end{array}$$

$i=1,2$ with f_i projective surjective flat morphisms of varieties over K_o having geometrically integral fibres, such that $Q(S_i) \subset K_i$ and such that there exist $L_i \in \mathrm{Pic}(X_i)$ ample relative to S_i such that $u_i^*(L_i) \simeq L$. Put $Z = S_1 \times_{K_o} S_2$ and consider the cartesian diagrams

$$\begin{array}{ccc} X^i & \xrightarrow{\;v_i\;} & X_i \\ \downarrow & & \downarrow \\ Z & \xrightarrow{\;p_i\;} & S_i \end{array}$$

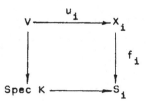

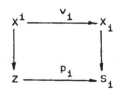

$i=1,2$ where p_i are as usual the canonical projections. Linear disjointness gives that the natural morphism $z:\mathrm{Spec}\,K \longrightarrow Z$ is dominant. Put $L^i = v_i^* L_i$ and apply Lemma (1.11) to this situation. Clearly $z \in Z'(K)$ with Z' as in (1.11) so Z' must contain a Zariski open subset U of Z. This immediately implies that there is some open set $S_2^* \subset S_2$ such that the closed fibres of $X_2 \times_{S_2} S_2^* \longrightarrow S_2^*$ are isomorphic as polarized varieties. We conclude by representability of <u>Isom</u> again (using once again the fact that isomorphisms preserving polarisations belong to finitely many components of the representing object for <u>Isom</u>).

(1.20) Last step is to prove (1.5)for arbitrary C. Put K=algebraic closure of $Q(S)$ and $V = X \times_S \mathrm{Spec}\,K$. By (1.1) we have that V is K-isomorphic to $V_o \otimes_{K_o} K$ where $K_o = K^{\Delta}$, $\Delta = \mathrm{Der}_C(\mathcal{O}_V)$ and V_o is some projective variety over K_o. Now look again at the diagram in (1.17) where C, K, K_o, V, V_o are those which we just defined. We claim that $\mathrm{Im}(\varphi) = \mathrm{Im}(\psi)$. The inclusion "$\subset$" is obvious; the inclusion "$\supset$" can be seen as follows. If $\delta \in \mathrm{Der}_{K_o}(K)$ then $\psi(\delta) = \varphi(\delta^*)$ where δ^* is the trivial lifting of δ to $V_o \otimes_{K_o} K = V$. A diagram chase shows that α must be injective. Now it is easy to express $V_o \longrightarrow \mathrm{Spec}\,K_o$ as the pull back of some $X'' \longrightarrow S''$ having the desired properties in (1.5)

Let us discuss some more consequences of our results.

(1.21) COROLLARY. Let V be a projective variety over an algebraically closed field K and let K_o be the smallest algebraically closed field of definition for V (1.3); write $V \simeq V_o \otimes_{K_o} K$ for some projective variety V_o over K_o (which is clearly unique up to K_o-isomorphism). Then the Kodaira-Spencer map $\mathrm{Der}(K_o) \longrightarrow \mathrm{Ext}^1(\Omega_{V_o/K_o}, \mathcal{O}_{V_o})$ is injective. In particular

$$\mathrm{tr.deg.}K_o/\mathbb{Q} \leq \dim_K \mathrm{Ext}^1(\Omega_{V/K}, \mathcal{O}_V)$$

Proof. Injectivity of the Kodaira-Spencer map follows from looking again at the diagram in (1.17) with $C=\mathbb{Q}$ and reasoning as in (1.20) above. Finally

$$\text{tr.deg.}K_0/\mathbb{Q} = \dim_{K_0}(\mathrm{Der}(K_0)) \leq \dim_{K_0} \mathrm{Ext}^1(\Omega_{V_0/K_0}, \mathcal{O}_{V_0}) =$$
$$= \dim_K \mathrm{Ext}^1(\Omega_{V/K}, \mathcal{O}_V)$$

(1.22) The following construction will be needed later. Let K be an algebraically closed Δ-field, V a projective Δ-variety over K, $C=K^{\Delta}$ and $\sigma: V \longrightarrow X \otimes_C K$ a descent isomorphism. Define a map

$$\hat{\sigma}: \Delta \longrightarrow \mathrm{Der}_K(\mathcal{O}_{X\otimes K}) = \mathrm{Der}_C(\mathcal{O}_X) \otimes_C K$$
$$\hat{\sigma}(\delta) = (\sigma^*)^{-1}\delta_V\sigma^* - \delta_K^*$$

where σ^* is the induced isomorphism $Q(X\otimes K) \longrightarrow Q(V)$ and δ_K^* is as usual the trivial lifting of δ_K to $X\otimes K$. Then let $\Lambda(\sigma)$ be the smallest Lie C-subalgebra of $\mathrm{Der}_C(\mathcal{O}_X)$ such that $\hat{\sigma}(\delta)$ is contained in $\Lambda(\sigma)\otimes_C K$ for all $\delta\in\Delta$. Note that even if all δ_V commute it may happen that $\Lambda(\sigma)$ does not have a commutative basis. Note also that σ is a splitting isomorphism if and only if $\Lambda(\sigma)=0$.

(1.23) COROLLARY. Let F/K be a Δ-function field with K algebraically closed. Suppose F/K has a projective Δ-model V with $\mathrm{Der}_K(\mathcal{O}_V)=0$ (i.e. with no non-zero global vector fields). Then V is split; in particular F/K is split.

Proof. Obvious from (1.1) and (1.22).

(1.24) COROLLARY. Let K be an algebraically closed Δ-field and V a projective Δ-variety. Then there exists a birational morphism of projective Δ-varieties $\tilde{V} \longrightarrow V$ with $\tilde{V}$ non-singular.

Proof. By (1.1) there is a descent isomorphism $\sigma: V \longrightarrow X\otimes K$. By

Hironaka's equivariant resolution (I.2.10) there is a birational morphism $\tilde{X} \longrightarrow X$ of projective varieties over C such that $\tilde{X}$ is non-singular and all vector fields on X lift to vector fields on $\tilde{X}$. Put $\tilde{V} = \tilde{X} \otimes K$ and let $D(\delta) \in \mathrm{Der}_C(\mathcal{O}_{\tilde{X}}) \otimes_C K$ be the lifting of $\hat{\sigma}(\delta) \in \mathrm{Der}_C(\mathcal{O}_X) \otimes K$. Then $\tilde{V}$ is a Δ-variety with $\delta_{\tilde{V}} = \delta_K^* + D(\delta)$ and $\tilde{V} \longrightarrow X \otimes K \xrightarrow{\sigma^{-1}} V$ is a birational Δ-morphism.

(1.25) Let K be an algebraically closed ordinary Δ-field and F/K a Δ-function field with no movable singularity and with tr. deg.$F/K = 1$, $\Delta = \{\delta\}$. In [Mtd] a fairly explicit description was given for such objects cf.pp.13,37,91. Let us show how one can reprove these structure results by using Theorem (1.1). Let V be the (unique) non-singular projective model of F/K. By hypothesis there is a projective Δ-model W. Since $W^{nor} = V$ we get by (I.2.2) that V is also a Δ-model, hence by (1.1) there is a descent isomorphism $\sigma: V \longrightarrow X \otimes_C K$ where $C = K^{\Delta}$ and X is a non-singular projective curve over C; put $D = \hat{\sigma}(\delta)$. There are three possibilities:

Case 1. X has genus 0, hence $X = \mathbb{P}^1_C$. If we put $Q(X) = C(t)$ with t an indeterminate then $\mathrm{Der}_C(\mathcal{O}_X)$ has a basis over C consisting of $\theta_0, \theta_1, \theta_2 \in \mathrm{Der}_C(Q(t))$ with $\theta_0 t = 1$, $\theta_1 t = t$, $\theta_2 t = t^2$, consequently $D = a_0\theta_0 + a_1\theta_1 + a_2\theta_2$, $a_0, a_1, a_2 \in K$ hence if we put $x = \sigma^* t$ we get $F = K(x)$ and $\delta x = a_0 + a_1 x + a_2 x^2$ which is precisely the Riccati equation (compare with [Mtd]p.13).

Case 2. X has genus 1, hence it is an elliptic curve. We may write $Q(X) = C(t,s)$ with $s^2 = P_3(t)$, P_3 being a polynomial of degree 3 with coefficients in C and with no multiple roots. Now $\mathrm{Der}_C(\mathcal{O}_X)$ has dimension 1 over C and a basis for it consists of the derivation θ for which $\theta t = s$. So $D = a\theta$ for some $a \in K$ and putting $x = \sigma^* t$ we get for $a \neq 0$ $F = K(x, \delta x)$ with $(\delta x)^2 = a^2 P_3(x)$ which is precisely the Weierstrass equation (compare with [Mtd]p.37). If $a = 0$ F/K is split.

Case 3. X has genus ≥ 2. In this case $\mathrm{Der}_K(\mathcal{O}_V) = 0$ and by (1.23) V is split (compare with [Mtd]p.91).

2. Splitting projective Δ-varieties.

The aim of this § is to prove the following:

(2.1) THEOREM. Let K be a Δ-field and V a projective Δ-variety over K. Then there exists a Δ-field extension K_1/K with $\text{tr.deg.}K_1/K < \infty$ and with K_1^{Δ}/K^{Δ} algebraic such that the Δ-variety $V \otimes_K K_1$ over K_1 is split. Moreover if V is a partial Δ-variety and K is algebraically closed one can take K_1/K to be strongly normal.

(2.2) COROLLARY. Let F/K be a Δ-function field with no movable singularity. Then there exists a Δ-field extension K_1/K with tr. $\text{deg.}K_1/K < \infty$ and with K_1^{Δ}/K^{Δ} algebraic such that $Q(F \otimes_K K_1)/K_1$ is split. Moreover if F is a partial Δ-field and K is algebraically closed then one can choose K_1/K to be strongly normal.

(2.3) Remark that Corollary (2.2) was proved in [Mtd] pp.30,48,91 in the case $\text{card}\,\Delta = \text{tr.deg.}F/K = 1$ by somewhat ad-hoc arguments based on special features of curves. Our proof will be more conceptual and will work in arbitrary dimension. The idea of our proof is to combine Theorem (1.1) with a stronger form of Kolchin's surjectivity theorem for the logarithmic derivative [Kol_1] pp.420-421 which will be proved below. Note also that Theorem (2.1) will play a key role in the next §.

(2.4) We start with a preparation. Let X be a projective variety over an algebraically closed field C and let K be a Δ-field with K^{Δ} containing C. Let $A=\text{Aut}_{X/C}$ the object representing the functor $\underline{\text{Aut}}_{X/C}=\underline{\text{Isom}}_{X/C,X/C}$; it is a group scheme locally of finite type over C having countably many connected components. We make the convention that A acts on X on the left so we have an action $\mu: A \times X \longrightarrow X$. The connected component of A will be denoted by $G=\text{Aut}^{\bullet}_{X/C}$; it is an algebraic group over C. Aut(X/C) and $\mathrm{Aut}^{\circ}(X/C)$ will denote the groups of C-points $A(C)=\mathrm{Aut}_{X/C}(C)$ and $G(C)=\mathrm{Aut}^{\circ}{}_{X/C}(C)$; we shall denote them sometimes simply by Aut(X), $\mathrm{Aut}^{\circ}(X)$. Now A(K) acts on $\mathrm{Der}_C(Q(X\otimes_C K))$ by the formula

$$(g,D)\longmapsto g^{**}D=(g^{*})^{-1}D\,g^{*}$$

where $g\in A(K)$, $D\in\mathrm{Der}_C(Q(X\otimes K))$ and g^{*} is the K-automorphism of $Q(X\otimes K)$ induced by g. Clearly if $D\in\mathrm{Der}_K(\mathcal{O}_{X\otimes K})=\mathrm{Der}_C(\mathcal{O}_X)\otimes_C K$ then $g^{**}D\in\mathrm{Der}_K(\mathcal{O}_{X\otimes K})$. Define

$$\log\delta : A(K)\longrightarrow \mathrm{Der}_C(\mathcal{O}_X)\otimes_C K \quad \text{by}$$

$$\log\delta\, g = g^{**}\delta_K{}^{*} - \delta_K{}^{*}$$

where $\delta_K{}^{*}$ is the trivial lifting of δ_K to $X\otimes K$ (clearly $\log\delta\, g$ vanishes on K and sends $\mathcal{O}_{X\otimes K}$ into itself so it is well defined). Clearly if F/K is a Δ-field extension then the following diagram is commutative:

$$\begin{array}{ccc} A(K) & \xrightarrow{\ \log\delta\ } & \mathrm{Der}_C(\mathcal{O}_X)\otimes_C K \\ \downarrow & & \downarrow \\ A(F) & \xrightarrow{\ \log\delta\ } & \mathrm{Der}_C(\mathcal{O}_X)\otimes_C F \end{array}$$

Note also that $\log\delta(A(C))=0$.

(2.5) Representability of $\underline{\mathrm{Aut}}_{X/C}$ immediately yelds an identification between the tangent space to G at the identity and $\mathrm{Der}_C(\mathcal{O}_X)$. On the other hand the above tangent space identifies with the Lie algebra $\mathrm{Lie}_C(G)$ of right invariant derivations of Q(G) vanishing on C. We obtain an identification $\mathrm{Lie}_C(G)\xrightarrow{\sim}\mathrm{Der}_C(\mathcal{O}_X)$. However we shall repeatedly need in what follows a more explicit identification between these two spaces which we now describe.

For any $D \in Lie_C(G)$ there is a commutative diagram

$$\begin{array}{ccc} \mathcal{O}_X & \xrightarrow{\mu^\#} & \mu_* \mathcal{O}_{G\times X} \\ \tilde{D} \downarrow & & \downarrow D\otimes 1 \\ \mathcal{O}_X & \xrightarrow{\mu^\#} & \mu_* \mathcal{O}_{G\times X} \end{array}$$

where μ is the action, $D\otimes 1$ is the unique derivation on $Q(G\times X)$ extending D and vanishing on $1\otimes Q(X)$ and $\tilde{D} \in Der_C(\mathcal{O}_X)$. This can be seen as follows. Let G act on $G\times X$

$$\begin{array}{ccc} \psi : G\times(G\times X) & \longrightarrow & G\times X \\ (g,(h,x)) & \longmapsto & (hg^{-1}, gx) \end{array}$$

We claim that $\mathcal{O}_X$ equals the sheaf $(\mu_* \mathcal{O}_{G\times X})^{(G,\psi)}$ of invariants of the action above in $\mu_* \mathcal{O}_{G\times X}$. Indeed if we let G act trivially on X , $t : G\times X \longrightarrow X$, $t(g,x)=x$ then $\mu : G\times X \longrightarrow X$ is a G-equivariant map with respect to the action of G on the source by ψ and on the target by t (we shall say that μ is (ψ,t)-equivariant). Now consider also the action

$$\begin{array}{ccc} \varphi : G\times(G\times X) & \longrightarrow & G\times X \\ (g,(h,x)) & \longmapsto & (hg^{-1}, x) \end{array}$$

One immediately checks that $\mu : G\times X \longrightarrow X$ is equivalent as a (ψ,t)-equivariant map with the second projection $p_2 : G\times X \longrightarrow X$ viewed as a (φ,t)-equivariant map: the equivariant isomorphism is given by $G\times X \longrightarrow G\times X$, $(g,x) \longmapsto (g, g^{-1}x)$.Now our claim follows since it is clear that $\mathcal{O}_X = (p_{2*}\mathcal{O}_{G\times X})^{(G,\varphi)}$ by Kunneth formula. Now to see that $(D\otimes 1)(\mathcal{O}_X) \subset \mathcal{O}_X$ it is sufficient to note that $D\otimes 1$ commutes with the action ψ. And indeed it commutes since multiplication by g in $G\times X$ is given by

$$G\times X \xrightarrow{R_{g^{-1}}\times 1} G\times X \xrightarrow{1\times g} G\times X$$

and D is right invariant. Now put $\tilde{D}$ = restriction of $D\otimes 1$ to $\mathcal{O}_X$ and we are done. The map $\lambda: \mathrm{Lie}_C(G) \longrightarrow \mathrm{Der}_C(\mathcal{O}_X)$, $\lambda(D)=\tilde{D}$ is of course an injective homomorphism of C-Lie algebras and since we already knew that $\dim_C \mathrm{Lie}_C(G) = \dim_C \mathrm{Der}_C(\mathcal{O}_X)$ it follows that λ is an isomorphism of Lie C-algebras

(2.6) The isomorphism λ defined above uniquely extends to an isomorphism of Lie K-algebras which will also be denoted by

$$\lambda: \mathrm{Lie}_K(G) \longrightarrow \mathrm{Der}_C(\mathcal{O}_X)\otimes_C K$$

Now both Lie K-algebras above are naturally equipped with derivations $\delta^{\#}$ $(\delta\in\Delta)$ defined as follows: $\delta^{\#}D=[\delta_{KG}^{*}, D]$, $\delta^{\#}d=[\delta_{KX}^{*}, d]$ where $D\in \mathrm{Lie}_K(G)$, $d\in \mathrm{Der}_C(\mathcal{O}_X)\otimes_C K$, δ_{KG}^{*} and δ_{KX}^{*} are the trivial liftings of δ_K to $G\otimes K$ and $X\otimes K$ respectively and $[\,,\,]$ is the Poisson bracket on $\mathrm{Der}_C(Q(G\otimes K))$ and $\mathrm{Der}_C(Q(X\otimes K))$ respectively. Then one immediately checks that λ agrees with this extra-structure in the sense that $\lambda(\delta^{\#}D)=\delta^{\#}(\lambda(D))$ for all D as above.

(2.7) The following easy remark will play a key role. We claim that in notations above there is a commutative diagram

$$\begin{array}{ccc} G(K) & \xrightarrow{\ \ell\delta\ } & \mathrm{Lie}_K(G) \\ \bigcap & & \downarrow \lambda \\ A(K) & \xrightarrow{\ \log\delta\ } & \mathrm{Der}_C(\mathcal{O}_X)\otimes_C K \end{array}$$

where $\ell\delta$ is Kolchin's logarithmic derivative (I.3.15). Indeed if we put $\mathcal{F}=\mathcal{O}_{X\otimes K}$ and $\mathcal{G}=\mu_{*}\mathcal{O}_{(G\times X)\otimes K}$ then for any $g\in G(K)$ we have a commutative diagram

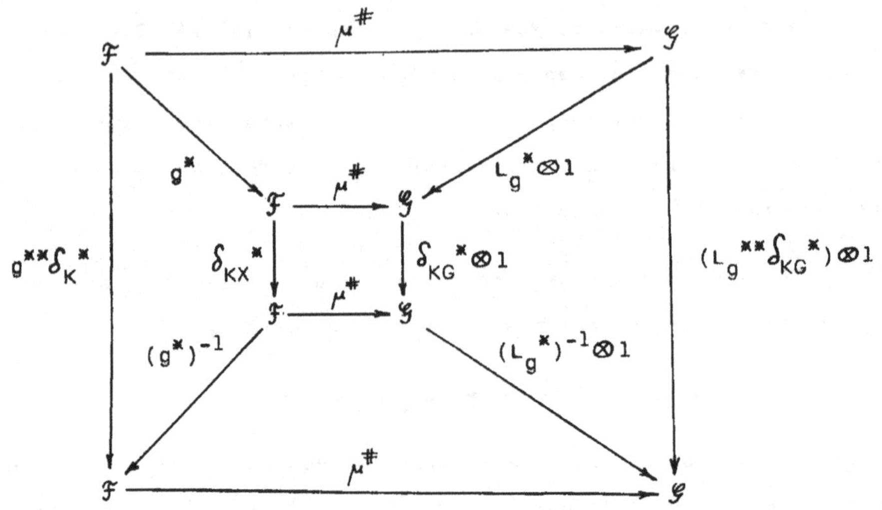

where $\delta_{KG}{}^* \otimes 1$ is the unique derivation extending $\delta_{KG}{}^*$ and vanishing on $Q(X \otimes K)$; same meaning for $(L_g{}^{**}\delta_{KG}{}^*) \otimes 1$. Now the two commutative squares give

$$(\delta_{KG}{}^* \otimes 1) \circ \mu^{\#} = \mu^{\#} \circ \delta_{KX}{}^* \quad \text{and}$$

$$(L_g{}^{**}\delta_{KG}{}^* \otimes 1) \circ \mu^{\#} = \mu^{\#} \circ (g^{**}\delta_{KX}{}^*)$$

Substracting and using (I.3.16) we get $((\ell\delta g) \otimes 1) \circ \mu^{\#} = \mu^{\#} \circ \log \delta g$; by (2.5) and (2.6) this equality reads $\lambda(\ell\delta g) = \log\delta g$ and we are done.

Now to prove Theorem (2.1) we need first to generalize Kolchin's theorem on the surjectivity of the logarithmic derivative $[Kol_1]$ pp. 420-421, [Kov] p.272 to the case when Δ is infinite and "non-commutative"; such a generalisation was done in [NW] p.982 for linear algebraic groups but even in this case we shall need a more precise statement. The following easy lemma will prove itself very useful here and later:

(2.8) LEMMA. Let K be a Δ-field, $C = K^{\Delta}$ being algebraically closed, let G be a connected algebraic group over C and $\varphi : \Delta \longrightarrow Lie_K(G)$ an arbitrary map. Let's give $G \otimes_C K$ a structure of Δ-variety by taking

$$\delta_{G\otimes K} = \delta_K^{*} + \varphi(\delta) \quad \text{for all} \quad \delta \in \Delta$$

where δ_K^{*} denotes as usual the trivial lifting of δ_K to $G\otimes K$. Let p be any point in $(G\otimes K)_\Delta$. Then the residual Δ-field extension $K(p)/K$ is a G-primitive extension with G-primitive given by

$$\alpha : \mathrm{Spec}\ K(p) \longrightarrow G\otimes K \longrightarrow G$$

and with $\ell\delta\alpha = \varphi(\delta)$ for all $\delta \in \Delta$.

Proof. It is just a matter of interpreting definitions ! Clearly $K(p)=K(\alpha)$ so we only have to check that $\ell\delta\alpha=\varphi(\delta)$ for all $\delta\in\Delta$. Now $\mathrm{Spec}\ K(p) \longrightarrow G\otimes K$ is a morphism of Δ-schemes over the Δ-scheme $\mathrm{Spec}\ K$ hence one gets a morphism of Δ-schemes $\alpha : \mathrm{Spec}\ K(p) \longrightarrow G\otimes K(p)$ and clearly $\delta_{G\otimes K(p)} = \delta_{K(p)}^{*} + \varphi(\delta)$ where $\delta_{K(p)}^{*}$ is the trivial lifting of $\delta_{K(p)}$ to $G\otimes K(p)$ and we still denoted by $\varphi(\delta)$ the image of $\varphi(\delta)$ in $\mathrm{Lie}_{K(p)}(G)$. We obtain a commutative diagram

$$\begin{array}{ccc}
\mathcal{O}_{G\otimes K(p),\alpha} & \xrightarrow{\ \delta_{K(p)}^{*} + \varphi(\delta)\ } & \mathcal{O}_{G\otimes K(p),\alpha} \\
\alpha^{*} \downarrow & & \downarrow \alpha^{*} \\
K(p) & \xrightarrow{\ \delta_{K(p)}\ } & K(p)
\end{array}$$

and we are done.

(2.9) COROLLARY. Let K, C, G and φ as in (2.8). Then:

1) There exists a Δ-field extension K_1/K such that K_1 is finitely generated over K as a field extension and $(K_1)^\Delta = C$ and there exists $\alpha\in G(K_1)$ such that $\ell\delta\alpha=\varphi(\delta)$ for all $\delta\in\Delta$.

2) Suppose K is a partial Δ-field, K^{+} is a constrainedly closed partial Δ-field extension of K and suppose that the following "in-

tegrability conditions" hold:

$$\delta^{\#}(\varphi(\theta)) - \theta^{\#}(\varphi(\delta)) + [\varphi(\delta),\varphi(\theta)] = 0$$

for all $\delta,\theta\in\Delta$. Then one can take K_1 from assertion 1) to be contained in K^+ and such that K_1/K is strongly normal with $G_{K_1/K}\subset G$.

Proof. 1) Apply Lemma (2.8) for any p which is maximal in $(G\otimes K)_\Delta$ and aplly 8) from (I.2.2). To prove 2) combine (I.3.18) and Lemma (2.8) by taking p to be the image of any Δ-morphism Spec $K^+ \longrightarrow G\otimes K$ over K (which exists by (I.3.10) since the integrability conditions imply that $G\otimes K$ is a partial Δ-scheme).

(2.10) Now we prove Theorem (2.1). Clearly we may suppose K algebraically closed. By Theorem (1.1) there is a descent isomorphism $\sigma: V\longrightarrow X\otimes_C K$ with $C=K^\Delta$. Put $G=\mathrm{Aut}^0(X/C)$ as in (2.4) and use notational conventions from (2.4)-(2.7) and (1.22). Define

$$\varphi:\Delta \xrightarrow{\hat{\sigma}} \mathrm{Der}_C(\mathcal{O}_X)\otimes_C K \xrightarrow{\lambda^{-1}} \mathrm{Lie}_K(G)$$

By 1) in (2.9) there is a Δ-field extension K_1/K of finite transcendence degree and with $(K_1)^\Delta=C$ and there exists $\alpha\in G(K_1)$ such that $\ell\delta\alpha=\varphi(\delta)$ for all $\delta\in\Delta$. If in addition we suppose that V is a partial Δ-variety then for all $\delta,\theta\in\Delta$ we have

$$0=\left[(\sigma^*)^{-1}\delta_V\sigma^*,(\sigma^*)^{-1}\theta_V\sigma^*\right]=\left[\delta_K{}^*+\hat{\sigma}(\delta),\theta_K{}^*+\hat{\sigma}(\theta)\right]=$$

$$=\delta^{\#}(\hat{\sigma}(\theta))-\theta^{\#}(\hat{\sigma}(\delta))+\left[\hat{\sigma}(\delta),\hat{\sigma}(\theta)\right]$$

and taking λ^{-1} we get

$$0=\delta^{\#}(\varphi(\theta))-\theta^{\#}(\varphi(\delta))+\left[\varphi(\delta),\varphi(\theta)\right]$$

Applying 2) from (2.9) we may suppose K_1/K is strongly normal. Now applying λ to $\ell\delta\alpha=\varphi(\delta)$ we get $\log\delta\alpha=\hat{\sigma}(\delta)$ for all $\delta\in\Delta$ cf.(2.7).

Let $\sigma_1=\sigma\otimes 1:V\otimes_K K_1 \longrightarrow X\otimes_C K_1$ be the K_1-isomorphism induced by σ and let $\delta_{K_1}{}^*$ be the trivial lifting of δ_{K_1} to $X\otimes_C K_1$. Then the image of $\hat{\sigma}(\delta)$ in $\mathrm{Der}_C(\mathcal{O}_X)\otimes_C K_1$ is

$$D = (\sigma_1{}^*)^{-1}\delta_{V\otimes K_1}\sigma_1{}^* - \delta_{K_1}{}^*$$

because D vanishes on $1\otimes K_1$ and its restriction to $Q(X)\otimes_C K$ equals $\hat{\sigma}(\delta)$. Since

$$\log\delta\alpha = (\alpha^*)^{-1}\,\delta_{K_1}{}^*\alpha^* - \delta_{K_1}{}^*$$

we get

$$(\beta^*)^{-1}\delta_{V\otimes K_1}\beta^* = \delta_{K_1}{}^*$$

where β is the isomorphism given by

$$\beta : V\otimes_K K_1 \xrightarrow{\ \sigma_1\ } X\otimes_C K_1 \xrightarrow{\ \alpha^{-1}\ } X\otimes_C K_1$$

and we are done.

3. Descent of Δ-points.

(3.1) Let K be an algebraically closed field and consider pairs (V,Σ) "over K" consisting of a variety V over K and a subset Σ of (not necessarily closed points of) V. For any $p\in\Sigma$ write

V_p=subvariety of V whose generic point is p

If C is an algebraically closed subfield of K we say that C is a field of definition for (V,Σ) (or that (V,Σ) descends to C,a.s.o) if there exists a pair (X,S) over C, a K-isomorphism $\sigma:V\longrightarrow X\otimes_C K$ and a bijection $\sigma^*:\Sigma\longrightarrow S$ such that

$$\sigma^*(X_{\sigma^*(p)} \otimes K) = V_p \quad \text{for all} \quad p \in \Sigma$$

Clearly $\sigma^\#$ is uniquely determined by σ. We shall write $\sigma:(V,\Sigma) \simeq (X,S) \otimes_C K$ and call σ a descent isomorphism for (V,Σ).

The aim of this § is to prove and discuss applications of the following :

(3.2) THEOREM. Let K be an algebraically closed Δ-field and V a projective Δ-variety over K. Then (V,V_Δ) descends to $C=K^\Delta$. More precisely there is a descent isomorphism

$$\sigma:(V,V_\Delta) \longrightarrow (X,X_\Lambda) \otimes_C K$$

where $\Lambda = \Lambda(\sigma)$ as defined in (1.22), X is viewed as a Λ-variety and X_Λ denotes as usual the set of Λ-points of X.

(3.3) First let's note that in order to prove the Theorem it is sufficient to check the first part of its conclusion namely that (V,V_Δ) descends to C. Indeed suppose that we have a descent isomorphism

$$\sigma:(V,V_\Delta) \longrightarrow (X,S) \otimes_C K$$

where S is a subset of X. We claim that $S=X_\Lambda$. Indeed if $a \in X_\Lambda$ then by definition of Λ the ideal I_{X_a} of the subvariety $X_a \subset X$ is stable under Λ. Since the ideal $I_{X_a \otimes K}$ of $X_a \otimes K \subset X \otimes K$ equals $I_{X_a} \otimes K$ it follows that $\sigma^* I_{X_a \otimes K}$ is a Δ-ideal hence $a \in S$. Conversely if $a \in S$ we have

$$((\sigma^*)^{-1} \delta_V \sigma^*)(I_{X_a} \otimes K) \subset I_{X_a} \otimes K \quad \text{for all} \quad \delta \in \Delta$$

Let $(f_j)_j$ be a basis of K/C and write $\hat{\sigma}(\delta) = \sum \lambda_j(\delta) \otimes f_j$ with $\lambda_j(\delta) \in \Lambda$. For any local section h of I_{X_a} and any $\delta \in \Delta$ we have:

$$\sum_j \lambda_j(\delta) h \otimes f_j = \hat{\sigma}(\delta)(h \otimes 1) = (\sigma^*)^{-1} \delta_V \sigma^*(h \otimes 1) = \sum_p h_p \otimes k_p$$

with h_p sections of I_{X_a} and $k_p \in K$. Writing $k_p = \sum c_{pj} f_j$ with $c_{pj} \in C$ we get

$$\lambda_j(\delta)h = \sum_p c_{pj} h_p$$

which is a local section of I_{X_a} hence I_{X_a} is stable under all $\lambda_j(\delta)$ hence under Λ and we conclude that $a \in X_\Lambda$.

(3.4) Next step in proving Theorem (3.2) is to check it under the assumption that V is split. More precisely we shall prove that if we are given a splitting isomorphism $\sigma: V \longrightarrow X \otimes_C K$ then σ induces a descent isomorphism $\sigma: (V, V_\Delta) \longrightarrow (X, X) \otimes_C K$. Indeed it is sufficient to prove the following general fact (see [MD_1], Appendix or [Bu_2]): if K is a Δ-field with $C = K^\Delta$, A is a C-algebra, δ_K^* is the trivial lifting of δ_K to $A \otimes_C K$ and if we view $A \otimes K$ as a Δ-ring with the "split structure" $\delta_{A\otimes K} = \delta_K^*$ then for any Δ-ideal J in $A \otimes K$ there is an ideal I in A such that $J = I \otimes K$. We recall the argument for completeness. If $I = J \cap (A \otimes 1)$, $J^* = I \otimes K$ and $J \setminus J^* \neq \emptyset$ then choose a basis $(e_k)_k$ of A as a C-vector space and take an element $a = \sum e_k \otimes a_k \in J \setminus J^*$, $a_k \in K$ for which the number of nonzero a_k's is minimal and such that at least one coefficient a_k equals 1. We have $\delta a = \sum e_k \otimes \delta a_k \in J$ for all $\delta \in \Delta$ hence $\delta a \in J^*$. Now there is an index k_1 and a derivation operator $\delta_1 \in \Delta$ such that $\delta_1 a_{k_1} \neq 0$; we get

$$a - a_{k_1}(\delta_1 a_{k_1})^{-1} \delta_1 a \in J^*$$

hence $a \in J^*$, contradiction. So we conclude that $J = J^*$ and we are done.

(3.5) Now let us complete the proof of Theorem (3.2). By Theorem (2.1) we may find a Δ-field extension K_1/K with $(K_1)^\Delta = C$ such that $V \otimes_K K_1$ is split over K_1. By (3.4) there is a descent isomorphism

$$\sigma: (V \otimes_K K_1, (V \otimes_K K_1)_\Delta) \longrightarrow (X, S) \otimes_C K_1$$

For any $p \in V_\Delta$ let $p_1 \in V \otimes_K K_1$ be the generic point of $V_p \otimes_K K_1$ and denote $X_{\sigma^{\#}(p_1)}$ simply by X_p. Put $\tilde{V} = X \otimes_C K$, $\tilde{V}_p = X_p \otimes_C K$ and let

$$F_p : \left\{\begin{matrix}\text{locally noetherian} \\ \text{schemes over } K\end{matrix}\right\} \longrightarrow \{\text{sets}\}$$

be the functor defined by

$$F_p(T) = \{ f \in \mathrm{Isom}_T(V \times T, \tilde{V} \times T) ;\ f^*(\tilde{V}_p \times T) = V_p \times T \}$$

where fibre products are taken over K. It is easy to check that the functors F_p are representable by closed subschemes R_p of the scheme R representing the functor $\underline{\mathrm{Isom}}_{V/K, \tilde{V}/K}$. Now we know that the intersection R^* of all R_p's is not empty (p running through V_Δ) because $\sigma : V \otimes_K K_1 \longrightarrow \tilde{V} \otimes_K K_1$ above corresponds to a K_1-point in $R(K_1)$ whose image lies in all R_p's. Since R^* is closed we have $R^*(K) \neq \phi$ which implies that (V, V_Δ) descends to C. Now we conclude by (3.3).

To formulate our first consequence let's make the following definition. A Δ-variety V will be called logarithmic at infinity if there exists a compactification $(V, \bar{V})$ of V (i.e. an open immersion of V into a projective variety $\bar{V}$) such that $\bar{V}$ is a Δ-variety and $\bar{V} \setminus V$ with its reduced scheme structure is a Δ-subscheme of $\bar{V}$.

(3.6) COROLLARY. Let K be an algebraically closed Δ-field and V a Δ-variety over K logarithmic at infinity. Then K^Δ is a field of definition for V.

Proof. It is clear from Theorem (3.2) and from the fact that if $(V, \bar{V})$ is a compactification with $\bar{V}$ a Δ-variety and $\bar{V} \setminus V$ a Δ-subscheme then all irreducible components of $\bar{V} \setminus V$ are Δ-subvarieties of $\bar{V}$, see (I.2.2).

(3.7) Let (V, Σ) be a pair over K as in (3.1). Then there is a natural structure of Δ-field on K for which V is a Δ-variety and $\Sigma\subset V_\Delta$ namely we put $\Delta=\Delta(V,\Sigma)$ with

$$\Delta(V,\Sigma)=\left\{\delta\in \mathrm{Der}(\mathcal{O}_V);\ \delta(K)\subset K \text{ and } \delta(m_p)\subset m_p \text{ for all } p\in\Sigma\right\}$$

Of course if V is projective then the condition $\delta(K)\subset K$ in the above definition is again superfluous. We have the following generalisation of (1.3) and of a result in [Bu_2]:

(3.8) COROLLARY. Suppose K is an algebraically closed field and (V,Σ) is a pair over K as above with V projective. Then the set of all algebraically closed fields of definition for (V,Σ) has a smallest element which equals $K^{\Delta(V,\Sigma)}$.

In what follows we apply (3.8) to discuss descent of "open varieties". There is a remarkable case in which things can be settled exactly as in the projective case, namely the case of "affine varieties with K^*-action":

(3.10) PROPOSITION. Let K be an algebraically closed field and $V=\mathrm{Spec}\ R$ an affine variety over K. Suppose R admits a structure of quasi-homogeneous K-algebra (i.e. $R=R_0\oplus R_1\oplus R_2\oplus\ldots$ is a graded K-algebra, $R_0=K$ and R is finitely generated over K). Then the set of all algebraically closed fields of definition for V has a smallest element which equals $K^{\Delta(V)}$.

Proof. Exactly as in (1.3) we see that is is sufficient to prove that $K^{\Delta(V)}$ is a field of definition for V. Put

$$\Delta(R,\mathrm{grad})=\left\{\delta\in \mathrm{Der}(R);\ \delta(R_n)\subset R_n \text{ for all } n\geq 0\right\}$$

We claim that the maps $\Delta(V)\longrightarrow \mathrm{Der}(K)$ and $\Delta(R,\mathrm{grad})\longrightarrow \mathrm{Der}(K)$ have the same image. Indeed let's take $\delta_V\in\Delta(V)$ and let δ_K be the restriction of δ_V to K. Let δ_R be the derivation on R induced by δ_V; we are looking for some $\tilde{\delta}_R$ whose restriction to K is still δ_K and which belongs to $\Delta(R, \text{grad})$. Let $p_n: R \longrightarrow R_n$ be the natural projections; then one checks that

$$\tilde{\delta}_R = \sum_{n \geq 0} p_n \delta_R p_n$$

is good for our purpose and the claim is proved. Now we see that the $\Delta(R,\text{grad})$-variety V is logarithmic at infinity since it admits the embedding $V \longrightarrow \bar{V} = \text{Proj } R[T]$ where $\deg(T)=1$ and $\delta T=0$ for all $\delta \in \Delta(R,\text{grad})$. Consequently by (3.6) V descends to $K^{\Delta(R,\text{grad})} = K^{\Delta(V)}$ and we are done.

(3.11) The following discussion will show that the descent technique developed in (II.2) and in the present § may sometimes replace the technique from (II.1).

Let V be a projective Δ-variety over K and $f: V \longrightarrow \mathbb{P}_K$ a morphism into a projective space. We say that (the Δ-variety) V is f-linear if there exists a structure of Δ-variety on $\mathbb{P}_K$ such that f is a Δ-morphism. In this case $W=f(V)$ will be a Δ-subvariety of $\mathbb{P}_K$ and by Theorem (3.2) applied to $(\mathbb{P}_K, (\mathbb{P}_K)_\Delta)$ it follows that there exists a K-automorphism σ of $\mathbb{P}_K$ and a subvariety $Y \subset \mathbb{P}_C$ over $C=K^\Delta$ such that $\sigma^*(Y \otimes_C K)=W$ in particular it follows that the homogenous ideal $I(W)$ of W in $\mathbb{P}_K$ has the property that $\sigma^* I(W)$ can be generated by polynomials with coefficients in C. Looking at the proof of (3.2) we see that the statement above can be proved independently of (II.1) hence making use only of logarithmic derivative technique in (II.2) and (II.3). Same remark holds for Proposition (3 .10) at least if R is homogenous i.e. if $R=R_0[R_1]$. Indeed going back to the notations there we see that we have a closed immersion $f: \bar{V} \rightarrow \mathbb{P}^N$, $N=\dim_K R_1$ given by representing $R[T]$ as a quotient of the symmetric algebra $S(R_0T+R_1)$ of R_0T+R_1. But now the structure of $\Delta(R, \text{grad})$-algebra of R lifts to a structure of $\Delta(R,\text{grad})$-algebra of $S(R_0T+R_1)$

hence $\overline{V}$ is f-linear. We conclude that (3.10) can be made independent of (II.1) at least in the homogenous case.

Another consequence of the remarks above is that if we are given a projective Δ-variety V and an f-linearisation with f a closed immersion then our method from (II.2)(II.3) is sufficient to prove that V descends to K^{Δ}. Remark on the other hand that it was proved in [MD_1] that if V is a smooth projective Δ-variety whose pluricanonical system $|\omega_{V/K}^{\otimes m}|$ is base point free for some $m \geqslant 1$ then V is f-linear where f is the m'th pluricanonical map. Putting these things together we obtain an alternative proof (avoiding (II.1)) for the following particular case of Theorem (1.1): let V be a non-singular projective Δ-variety over K with ample canonical bundle; then V descends to K^{Δ}.

Note however that there is no f-linearisation (with f a closed immersion) for general projective Δ-varieties: look for instance at abelian varieties with derivations given by global vector fields. Consequently (II.1) cannot be avoided in general.

(3.12) Coming back to the descent of open varieties we should say that the problem whether any variety V (over an algebraically closed field K) has a smallest algebraically closed field of definition seems to be open. If such a field exists then it is easy to see that it must be equal to $K^{\Delta(V,\log)}$ where

$$\Delta(V,\log)=\left\{\begin{array}{l|l}\delta\in \mathrm{Der}(\mathcal{O}_V) & V \text{ viewed as a } \{\delta\}\text{-variety is} \\ \delta(K)\subset K & \text{logarithmic at infinity}\end{array}\right\}$$

On the other hand one has the following easy result:

(3.13) PROPOSITION. Let K be algebraically closed, V a non-singular variety over K and suppose there exists a compactification $(V,\overline{V})$ with the following properties:

1) $\overline{V}$ is normal

2) $D=\overline{V}\setminus V$ is of pure codimension 1

3) for any compactification $(V,\tilde{V})$ with $\tilde{V}$ non-singular and $\tilde{V}\setminus V$ a divisor with normal crossings the rational map $\tilde{V}\dashrightarrow\overline{V}$ is everywhere defined.

Then putting

$$\Delta(V,\overline{V})=\{\delta\in \mathrm{Der}(\mathcal{O}_{\overline{V}});\ \delta(I_D)\subset I_D\}$$

where I_D is the reduced ideal sheaf of D, we have that the set of all algebraically closed fields of definition for V has a smallest element which equals $K^{\Delta(V,\overline{V})}$.

Proof. V is a $\Delta(V,\overline{V})$-variety logarithmic at infinity so $K^{\Delta(V,\overline{V})}$ is a field of definition for V by Theorem (3.6). Now let C be an algebraically closed field of definition for V so $V\simeq X\otimes_C K$ for some non-singular variety X over C. By Hironaka's work there is a compactification $(X,\overline{X})$ of X with $\overline{X}$ non-singular and $\overline{X}\setminus X$ a divisor with normal crossings. Put $\tilde{V}=\overline{X}\otimes_C K$; then $\tilde{V}\setminus V$ is a divisor with normal crossings and by 3) the rational map $f:\tilde{V}\dashrightarrow\overline{V}$ is everywhere defined. Since $\overline{V}$ is normal, $f_*\mathcal{O}_{\tilde{V}}=\mathcal{O}_{\overline{V}}$. Now take any $\delta\in \mathrm{Der}_C(K)$ and let's prove that δ belongs to the image of $\Delta(V,\overline{V})\longrightarrow \mathrm{Der}(K)$. Let $\delta^*\in \mathrm{Der}(\mathcal{O}_{\tilde{V}})$ be the trivial lifting of δ to $\tilde{V}$. By (I.2.5) $\delta^*\in \mathrm{Der}(\mathcal{O}_{\overline{V}})$. Let $p\in\overline{V}$ be the generic point of any irreducible component of $\overline{V}\setminus V$. Since $\overline{V}$ is normal and since p has codimension 1 we get that $f^{-1}(p)$ consists of one point only ; denote it by q. We have $q\in\tilde{V}\setminus V=(\overline{X}\setminus X)\otimes K$ and $\mathcal{O}_{\overline{V},p}=\mathcal{O}_{\tilde{V},q}$. Hence q is a generic point of a component of the divisor $(\overline{X}\setminus X)\otimes K$ which clearly is a $\{\delta^*\}$-subvariety We conclude that δ^* takes the maximal ideal of $\mathcal{O}_{\overline{V},p}$ into itself hence $\delta^*\in\Delta(V,\overline{V})$ and we are done.

The above Proposition suggests that fields of definition of open varieties are related to the existence of "minimal compactifications". It would be interesting to get a better understanding of this relation. Of course it is easy to give examples of varieties V admitting a compactification $(V,\overline{V})$ as in (3.13) (take V any affine piece of curve or of a minimal non-ruled surface) and examples of V which don't ($V=\mathbb{P}^1\times W$ with W an affine curve).

(3.14) A difficult problem seems to be the existence and computation of the smallest algebraically closed field of definition of a function field F/K (see the definition below); this problem seems to be related to understanding "minimal models" of F/K.

Given a finitely generated field extension F/K with K algebraically closed we say that a subfield C of K is a field of definition for F/K if there exists a subfield D of F containing C such that a) F=KD and b) D and K are linearly disjoint over $C=K\cap D$, (or equivalently if there exists a variety X over C such that $Q(X\otimes_C K)$ is K-isomorphic to F). Let's consider the problem whether the set of all algebraically closed fields of definition for F/K has a smallest element and also the problem of computing this element. These problems are solved if tr.deg.F/K=1 by Corollary (1.3). We can solve these problems in three more cases:

tr.deg.F/K=2	(surface case)
tr.deg.F/K=$\varkappa$(F/K)	(general type case)
tr.deg.F/K=α(F/K)	

where $\varkappa$(F/K) is the Kodaira dimension of any non-singular projective model V of F/K [U] p.68 and α(F/K) is the dimension of the image of $V\longrightarrow \mathrm{Alb}(V)$.

(3.15) PROPOSITION. Let K be an algebraically closed field and F a finitely generated field extension of K of transcendence degree 2.

Let V be a non-singular projective model of F/K and put

$$V_{min}=\begin{cases}\text{a point, if } V \text{ is rational}\\ Alb(V)\text{, if } V \text{ is irrationally ruled}\\ \text{the minimal model of } V\text{, if } V \text{ is non-ruled}\end{cases}$$

Then the set of all algebraically closed fields of definition of F/K has a smallest element wiich equals $K^{\Delta(V_{min})}$.

Proof. It is an immediate consequence of Theorem (1.1) and Remark (1.8)

To formulate our next result note that for any finitely generated field extension F/K with K algebraically closed we have a birational invariant $R=R(F/K)$ called the canonical ring defined as

$$R=\sum_{k\geq 0} R_k, \qquad R_k=H^0(V,\omega_{V/K}^{\otimes k})$$

where V is any non-singular projective model. It is defined up to graded K-isomorphisms [U] p.66. Put

$$\Delta(F/K,can)=\{\delta\in Der(R),\ \ \delta(R_k)\subset R_k \ \text{ for all } \ k\geq 0\}$$

Then the subfield $K^{\Delta(F/K,can)}$ of K is well defined.

(3.16) PROPOSTTION. Let F/K be a finitely generated field extension with K algebraically closed. Suppose it is of general type. Then the set of all algebraically closed fields of definition for F/K has a smallest element which equals $K^{\Delta(F/K,can)}$.

Proof. Let C be an algebraically closed field of definition for F/K so $F\simeq Q(X\otimes_C K)$ over K for some variety X over C. We may suppose X is non-singular and projective. Then $R(F/K)\simeq R(Q(X)/C)\otimes_C K$ hence any $\delta\in Der_C(K)$ will belong to $Im(\Delta(F/K,can)\longrightarrow Der(K))$ so $K^{\Delta(F/K,can)}\subset C$. To see that $K^{\Delta(F/K,can)}$ is a field of definition for

F/K choose any non-singular projective model V of F/K and an integer n such that the n'th pluricanonical map $\varphi_n : V \longrightarrow \mathbb{P} = \mathbb{P}(H^0(V, \omega_{V/K}^{\otimes n})^{\vee})$ is birational onto its image cf. [U] p.56. The closure V_n of the image of φ_n identifies with Proj R^* where R^* is the K-subalgebra of R generated by R_n while $\mathbb{P}$=proj $S(R_n)$ where $S(R_n)$ is the symmetric algebra of R_n. Now $\delta(R^*) \subset R^*$ for all $\delta \in \Delta(F/K,can)$ while $S(R_n)$ has a natural srructure of $\Delta(F/K,can)$-ring induced by the maps $\delta : R_n \longrightarrow R_n$. Moreover the natural surjection $p : S(R_n) \longrightarrow R^*$ is a $\Delta(F/K,can)$-morphism hence Ker(p) is a $\Delta(F/K,$ can)-ideal, hence V_n is a $\Delta(F/K,can)$-subvariety of $\mathbb{P}$. By (3.2) V_n descends to $K^{\Delta(F/K,can)}$ and we are done. Note that exactly as in (3.11) one can make this proof independent of (II.1).

(3.17) PROPOSITION. Let K be an algebraically closed field, F a finitely generated field extension of K and suppose F/K has a projective normal model V such that for any non-singular projective model $\tilde{V}$ the rational map $\tilde{V} \dashrightarrow V$ is everywhere defined. Then the set of all algebraically closed fields of definition for F/K has a smallest element which equals $K^{\Delta(V)}$. In particular this happens if tr.deg.F/K= $=\alpha(F/K)$.

Proof. The first statement is an immediate consequence of Theorem (1.1) and Remark (1.8) and Hironaka's resolution of singularities. Now if tr. deg.F/K$=\alpha(F/K)$ we can take any non-singular projective model W of F/K and let V be the normalisation of Im(W $\longrightarrow$ Alb(W)) in F.

(3.18) We close this § by discussing a different type of applications of Theorem (3.2) namely applications related to "solutions of Δ-systems" cf.(IV.2). Let V be a Δ-variety over K and define

$$V_C = \{p \in V_\Delta : [K(p)^\Delta : K^\Delta] < \infty\}$$

As will become clear in (IV.2) any "solution" yelds a point in V_C and this is what justifies our interest below in determining V_Δ and especially V_C for a given V. Now note that what Theorem (3.2) says is that determination of V_Δ is reduced to determination of X_Λ which definitely is a more geometric object since derivations in Λ are global vector fields on X and X_Λ is the set of all generic points of irreducible "algebraic integral subvarieties" for these vector fields. We shall consider for simplicity the case when K is algebraically closed. The case of Δ-systems in (IV.2) can be reduced to this case by base change. So for instance when we shall prove that for a certain V the set V_C consists of closed points only this will translate into the fact that the corresponding Δ-systems "covering" V have only algebraic solutions in the sense of (IV.2.4)

(3.19) PROPOSITION. Let V be a split projective Δ-variety over an algebraically closed Δ-field K and let $\sigma : V \longrightarrow X \otimes_C K$ be a splitting isomorphism, $C = K^\Delta$. Then V_Δ identifies set-theoretically via σ with X while V_C identifies with the set of closed points of X. In particular V_C consists of closed points only.

Proof. Apply (3.4) plus the fact that if $p \in V_\Delta$ then V_p is still a split Δ-variety.

(3.20) PROPOSITION. Let V be a Δ-variety over an algebraically closed Δ-field and suppose V is an abelian variety. The following are equivalent:

1) V is split.

2) V_Δ contains a closed point of V.

3) V_Δ contains the generic point of some irreducible ample divisor.

Proof. 1)$\Longrightarrow$3) follows from (3.19). Now to prove the remaining implications apply Theorem (3.2) to get $(V, V_\Delta) \simeq (X, X_\Lambda) \otimes_C K$; by [Mum]

p. 124 X must be an abelian variety too. Let's prove 2)$\Longrightarrow$1). By hypothesis X_Λ must contain a closed point of X in other words the global vector fields in Λ must vanish at some point of X; consequently they must vanish everywhere so $\Lambda=0$ and V is split. Let's prove 3)$\Longrightarrow$2). There exists an ample irreducible divisor Y on X which is a Λ-subvariety of X. Since Λ consists of vector fields invariant under translations, $Y+x$ is a Λ-subvariety of X for all C-points $x\in X(C)$. Now choosing sufficiently general points $x_1,\dots,x_n\in X(C)$ where $n=\dim X$ we get by (I.2.2) that $Z=(Y+x_1)\cap\dots\cap(Y+x_n)$ is a (non-empty !) finite set of closed points of X all belonging to X_Λ which closes the proof.

(3.21) PROPOSITION. Let V be a Δ-variety over an algebraically closed Δ-field and suppose V is an abelian variety. The following are equivalent:

1) V_C contains the generic point of V (i.e. $Q(V)^\Delta=K^\Delta$).

2) V_Δ consists of the generic point of V only.

3) V_Δ does not contain codimension one points of V.

Proof. 2)$\Longrightarrow$3) is trivial; 3)$\Longrightarrow$1) follows from 7) in (I.2.2). Let's prove 1)$\Longrightarrow$3). Suppose there is a codimension one point of V in V_Δ; denote it by p and let W be the irrducible divisor corresponding to it. By [Ni] for instance, there is a surjective morphism of abelian varieties $f:V\longrightarrow V^*$ and an ample irreducible divisor W^* on V^* such that $\dim V^*>0$, $\mathrm{Ker}(f)$ is connected and $f^*W^*=W$. By 2) in (I.2.5) and 1) in (I.2.4) V^* is a Δ-variety and W^* is a Δ-subvariety of V^*. By (3.2o) V^* is split. Since $K^\Delta\subset Q(V^*)^\Delta\subset Q(V)^\Delta=K^\Delta$ we get $Q(V^*)=K$ contradiction; hence 1)$\Longrightarrow$3) is proved.

Let's prove 3)$\Longrightarrow$2). We shall prove that V_Δ does not contain codimension k points ($k\geqslant 1$) by induction on k. Step $k=1$ is precisely 3). Now let's make the induction step. Write again $(V,V_\Delta)\simeq(X,X_\Lambda)\otimes K$

as in the proof of (3.20) and suppose that V_Δ contains a codimension k Δ-point. Hence X contains a codimension k Λ-point; in other words X contains a codimension k Λ-subvariety Y. Choose any complete curve $Z \subset X$ not contained in Y and let Y^* be the image of the map $Y \times Z \longrightarrow X$, $(y,z) \longmapsto y+z$. Then Y^* is a codimension k-1 Λ-subvariety of X implying that V_Δ contains a codimension k-1 point, contradiction.

(3.22) The last example we are going to consider will be the projective space $\mathbb{P}_K^N$ (K algebraically closed) as a Δ-variety. Applying Theorem (3.2) we may write $(\mathbb{P}_N^N,(\mathbb{P}_K^N)_\Delta) \simeq (\mathbb{P}_C^N,(\mathbb{P}_C^N)_\Lambda) \otimes_C K$ where $C=K^\Delta$ and Λ is a Lie sub-C-algebra of the Lie C-algebra $sl(N+1,C)=$ $=Lie_C(Aut(\mathbb{P}_C^N/C))$.

Let's start with the simplest case N=1 (as already noted in (1.22) this is the Riccati equation). One immediately checks that in this case $(\mathbb{P}_C^1)_\Lambda$ has at most 2 closed points unless $\Lambda=0$. We deduce that $(\mathbb{P}_K^1)_\Delta$ has at most 2 closed points unless $\mathbb{P}_K^1$ is split.

Let's consider the case N =2. If $\Lambda=0$, $\mathbb{P}_K^N$ is of course split. If $\dim \Lambda =1$ the determination of $(\mathbb{P}_C^2)_\Lambda$ is a classically solved problem going back to Darboux, see [J_1]pp.8-19. If $\dim \Lambda \geqslant 2$ the analysis is trivial modulo the case $\dim \Lambda =1$.

Finally let's say a few words about the case $N \geqslant 3$. The problem of describing $(\mathbb{P}_C^N)_\Lambda$ here becomes quite complicated and not much is known in the general case ; see however [J_1], [Li_1]. It worths adding the following remark. Suppose $W \subset \mathbb{P}_K^N$ is a non-singular codimension one Δ-subvariety and suppose $tr.deg.Q(\mathbb{P}_K^N)^\Delta/K^\Delta \leqslant N-1$ (the latter assumption can be interpreted as follows: there are at most N-1 "independent" algebraic prime integrals cf. (IV.2)). Then the hypersurface W has degree at most 2. Indeed let $d=\deg W$ and let $f(x_0,\dots,x_N)=0$ be the equation of W. Then consider the subvariety $V^* \subset \mathbb{P}_K^{N+1}$ given by the equation

$$x_{N+1}{}^{d}-f(x_0,\dots,x_N)=0$$

By (I.2.8) V^* has a natural structure of Δ-variety such that the projection

$$V^* \longrightarrow \mathbb{P}_K{}^N$$

$$(x_0,\dots,x_N,x_{N+1}) \longmapsto (x_0,\dots,x_N)$$

is a Δ-morphism. Now V^* is a smooth hypersurface in $\mathbb{P}_K{}^{N+1}$ of degree d and if $d \geqslant 3$ then $Der_K(\mathcal{O}_{V^*})=0$ [KS] hence by (1.23) $Q(V^*)/K$ issplit. On the other hand by (I.1.2) the extension $Q(V^*)^{\Delta}/Q(\mathbb{P}_K{}^N)^{\Delta}$ is algebraic so we get $tr.deg. Q(V^*)^{\Delta}/K^{\Delta} \leqslant N-1$ consequently $tr.deg. Q(V^*)/K \leqslant N-1$, contradiction. We conclude that $d \leqslant 2$ as claimed above.

4. Descent of Δ-Galois group.

(4.1) Let V_1 and V_2 be two varieties over a field K. We denote by $Rat(V_1,V_2)$ the set of all rational maps $V_1 \dashrightarrow V_2$ over K. If C is a subfield of K and Σ is a subset of $Rat(V_1,V_2)$ we say that C is a field of definition for (V_1,V_2,Σ) if there exists a triple (X_1,X_2,S) consisting of varieties X_1,X_2 over C and a set $S \subset Rat(X_1,X_2)$ and if there exist K-isomorphisms $\varphi_i : V_i \longrightarrow X_i \otimes_C K$, $i=1,2$ and a bijection $\varphi^{\#}:\Sigma \longrightarrow S$ such that for any $f \in \Sigma$ we have:

$$(\varphi_1 \times \varphi_2)^*(\Gamma_{\varphi^{\#}(f)} \otimes K) = \Gamma_f$$

where $\varphi_1 \times \varphi_2 : V_1 \times V_2 \longrightarrow (X_1 \times X_2) \otimes K$ and $\Gamma_f \subset V_1 \times V_2$, $\Gamma_{\varphi^{\#}(f)} \subset X_1 \times X_2$ are the closures of the graphs of f and $\varphi^{\#}(f)$ respectively. Here $V_1 \times V_2$ and $X_1 \times X_2$ stand for $V_1 \times_K V_2$ and $X_1 \times_C X_2$ respectively.

Note that if f is everywhere defined, the projection $\Gamma_f \longrightarrow V_1$ is an isomorphism hence so is $\Gamma_{\varphi^{\#}(f)} \longrightarrow X_1$ hence $\varphi^{\#}(f)$ is also everywhere defined and we have a commutative diagram:

$$\begin{array}{ccc} V_1 & \xrightarrow{\ f\ } & V_2 \\ \varphi_1 \downarrow & & \downarrow \varphi_2 \\ X_1 \otimes K & \xrightarrow[\varphi^{\#}(f)\otimes 1]{} & X_2 \otimes K \end{array}$$

Similarily if f is a rational dominant map (respectively a birational map) then so is $\varphi^{\#}(f)$ and we have a commutative diagram:

$$\begin{array}{ccc} Q(V_2) & \xrightarrow{\ f^{*}\ } & Q(V_1) \\ \varphi_2^{*} \uparrow & & \uparrow \varphi_1^{*} \\ Q(X_2 \otimes K) & \xrightarrow[\varphi^{\#}(f)^{*}\otimes 1]{} & Q(X_1 \otimes K) \end{array}$$

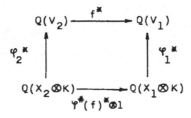

Now if V_1 and V_2 are Δ-varieties we denote by $Rat_\Delta(V_1, V_2)$ the set of all Δ-rational maps from V_1 to V_2 cf. (I.1.2). The aim of the present § is to prove and give applications of the following:

(4.2) THEOREM. Let K be an algebraically closed Δ-field and V_1 V_2 projective Δ-varieties over K. Then K^Δ is a field of definition for $(V_1, V_2, Rat_\Delta(V_1, V_2))$.

Proof. By Theorem (1.1) there exist descent isomorphisms

$$\sigma_i : V_i \longrightarrow X_i \otimes_C K, \ i=1,2, \ C=K^\Delta .$$

By Theorem (3.2) there exists a descent isomorphism

$$\sigma : (V_1 \times V_2, (V_1 \times V_2)_\Delta) \longrightarrow (X, S) \otimes_C K$$

where $S \subset X$. By representability of $\underline{Isom}_{X_1 \times X_2/C, X/C}$ there exists a

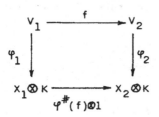

C-isomorphism $\varepsilon : X \longrightarrow X_1 \times X_2$. We claim there exists a commutative diagram

$$\begin{array}{ccccc} V_1 \times V_2 & \xrightarrow{\sigma} & X \otimes K & \xrightarrow{\varepsilon \otimes 1} & (X_1 \times X_2) \otimes K \\ \downarrow{\scriptstyle \sigma_1 \times \sigma_2} & & & & \uparrow{\scriptstyle \rho \otimes 1} \\ (X_1 \times X_2) \otimes K & & \xrightarrow[\tau_1 \times \tau_2]{} & & (X_1 \times X_2) \otimes K \end{array}$$

with $\rho \in \mathrm{Aut}(X_1 \times X_2/C)$ and $\tau_i \in \mathrm{Aut}(X_i \otimes K/K)$ for $i=1,2$. Indeed put $\tau = (\varepsilon \otimes 1)\sigma(\sigma_1 \times \sigma_2)^{-1}$ and let $A = \mathrm{Aut}_{X_1 \times X_2/C}$, $G = \mathrm{Aut}^{\circ}_{X_1 \times X_2/C}$ as in (2.4). Then τ identifies with a K-point in the group $A(K)$. Since $A(K)/G(K)$ identifies with $A(C)/G(C)$ we may write $\tau = (\rho \otimes 1)\tau_{\circ}$ with $\rho \in A(C)$ and $\tau_{\circ} \in G(K)$. But now the natural injective map

$$\mathrm{Aut}^{\circ}_{X_1/C} \times \mathrm{Aut}^{\circ}_{X_2/C} \longrightarrow \mathrm{Aut}^{\circ}_{X_1 \times X_2/C}$$

is an isomorphism just because its tangent map at the identity

$$\mathrm{Der}_C(\mathcal{O}_{X_1}) \oplus \mathrm{Der}_C(\mathcal{O}_{X_2}) \longrightarrow \mathrm{Der}_C(\mathcal{O}_{X_1 \times X_2})$$

is an isomorphism. Hence $\tau_{\circ} = \tau_1 \times \tau_2$ with $\tau_i \in \mathrm{Aut}^{\circ}_{X_i/C}(K) = \mathrm{Aut}^{\circ}(X_i \otimes K/K)$ and our claim is proved.

Put $\varphi_i = \tau_i \sigma_i$ for $i=1,2$. Now it is easy to check that for any $f \in \mathrm{Rat}_{\Delta}(V_1, V_2)$, Γ_f is a Δ-subvariety of $V_1 \times V_2$ hence there is a subvariety $Z_f \subset X$ such that $\sigma^*(Z_f \otimes K) = \Gamma_f$. Putting $Y_f = \rho^{-1}\varepsilon(Z_f)$ we have

$$(\varphi_1 \times \varphi_2)^*(Y_f \otimes K) = \Gamma_f$$

One immediately sees that $Y_f \longrightarrow X_1$ is a birational morphism hence $Y_f = \Gamma_{\varphi^{\#}(f)}$ for some $\varphi^{\#}(f) \in \mathrm{Rat}(X_1, X_2)$ and we are done.

(4.3) COROLLARY. Let K be an algebraically closed field, V_1, V_2 projective varieties over K and Σ a subset of $\mathrm{Rat}(V_1, V_2)$. Let

$\Delta(V_1,V_2,\Sigma)$ be the set of all pairs $(\delta_1,\delta_2)\in \mathrm{Der}(\mathcal{O}_{V_1})\times_{\mathrm{Der}(K)}\mathrm{Der}(\mathcal{O}_{V_2})$ having the property that for any $f\in\Sigma$ the map $f:V_1 \dashrightarrow V_2$ is a Δ-rational map with respect to the structures of ordinary Δ-varieties on V_1 and V_2 given by δ_1 and δ_2 respectively. Then the set of all algebraically closed fields of definition for (V_1,V_2,Σ) has a smallest element which equals $K^{\Delta(V_1,V_2,\Sigma)}$.

Proof. It works exactly as for (1.3).

Here comes our main application of Theorem (4.2); it says roughly speaking that Δ-Galois groups of Δ-function fields with no movable singularity "descend to constants":

(4.4) THEOREM. Let F/K be a Δ-function field with no movable singularity and with K algebraically closed. Then there exists a finitely generated field extension D of $C=K^{\Delta}$, a subset Λ of $\mathrm{Der}_C(D)$, a K-isomorphism $\psi^{*}:Q(D\otimes_C K)\longrightarrow F$ and a group isomorphism $\psi^{\#}$: $\mathrm{Gal}_{\Delta}(F/K)\longrightarrow \mathrm{Gal}_{\Lambda}(D/C)$ such that for any $\sigma\in \mathrm{Gal}_{\Delta}(F/K)$ the following diagram is commutative:

$$\begin{array}{ccc} F & \xrightarrow{\ \sigma\ } & F \\ \uparrow{\scriptstyle \psi^{*}} & & \uparrow{\scriptstyle \psi^{*}} \\ Q(D\otimes_C K) & \xrightarrow[\psi^{\#}(\sigma)\otimes 1]{} & Q(D\otimes_C K) \end{array}$$

Moreover given a projective Δ-model V of F/K one can find a descent isomorphism $\psi:V\longrightarrow X\otimes_C K$ such that one can take $D=Q(X)$, $\Lambda=\Lambda(\psi)$ and ψ^{*}=field isomorphism induced by ψ.

Remark. Clearly $\psi^{\#}$ is uniquely determined by ψ and even by ψ^{*}; such a ψ or ψ^{*} will be called a descent isomorphism for $\mathrm{Gal}_{\Delta}(F/K)$.

Proof. Fix a projective Δ-model V. By Theorem (4.2) $(V,V,Rat_{\Delta}(V,V))$ descends to C; let $\varphi_i : V \longrightarrow X\otimes_C K$ $i=1,2$ and $\varphi^{\#}: Rat_{\Delta}(V,V) \longrightarrow S \subset Rat(X,X)$ be the isomorphisms given by (4.2). Clearly $\varphi^{\#}$ takes the group $Bir_{\Delta}(V/K)$ of invertible elements in $Rat_{\Delta}(V,V)$ into some subset of the group $Bir(X/C)$ of invertible elements in $Rat(X,X)$. Now the natural maps $Bir_{\Delta}(V/K) \longrightarrow Gal_{\Delta}(F/K)$, $f \longmapsto f^*$ and $Bir(X/C) \longrightarrow Gal(D/C)$, $g \longmapsto g^*$ are isomorphisms so $\varphi^{\#}$ induces an injective map (which is not in general a group homomorphism) still denoted by $\varphi^{\#}: Gal_{\Delta}(F/K) \longrightarrow Gal(D/C)$ such that for any $\sigma \in Gal_{\Delta}(F/K)$ we have a commutative diagram

$$\begin{array}{ccc} F & \xrightarrow{\quad\sigma\quad} & F \\ {\scriptstyle \varphi_2^*}\big\uparrow & & \big\uparrow{\scriptstyle \varphi_1^*} \\ Q(D\otimes_C K) & \xrightarrow[\varphi^{\#}(\sigma)\otimes 1]{} & Q(D\otimes_C K) \end{array}$$

In particular we have $\varphi^{\#}(1)\otimes 1 = (\varphi_1^*)^{-1}\varphi_2^*$. Define $\psi^{\#}: Gal_{\Delta}(F/K) \longrightarrow Gal(D/C)$ by the formula $\psi^{\#}(\sigma)=\varphi^{\#}(\sigma)(\varphi^{\#}(1))^{-1}$ and put $\psi=\varphi_1$. Then for any $\sigma \in Gal_{\Delta}(F/K)$ the diagram in the statement of Theorem (4.4) is commutative and one immediately checks that $\psi^{\#}$ is a group homomorphism. All we have to prove now is that $Im(\psi^{\#})=Gal_{\Lambda}(D/C)$ with $\Lambda=\Lambda(\psi)$=smallest Lie C-subalgebra of $Der_C(\mathcal{O}_X)$ such that $\hat{\psi}(\delta) \in \Lambda(\psi)\otimes K$ for all $\delta \in \Delta$. Take $g \in Gal(D/C)$; we have $g \in Im(\psi^{\#})$ if and only if $g\otimes 1$ commutes with $(\psi^*)^{-1}\delta_V\psi^*$ for all $\delta\in\Delta$. Since $g\otimes 1$ clearly commutes with δ_K^* it follows that $g \in Im(\psi^{\#})$ if and only if $g\otimes 1$ commutes with $\hat{\psi}(\delta)$ for all $\delta\in\Delta$. Now if $g \in Gal_{\Lambda}(D/C)$ then $\hat{\psi}(\delta)$ clearly commutes with $g\otimes 1$. Conversely if $g\otimes 1$ commutes with all $\hat{\psi}(\delta)$'s choose a basis $(f_j)_j$ of K/C and write $\hat{\psi}(\delta)=\sum \lambda_j(\delta)\otimes f_j$ with $\lambda_j(\delta)\in\Lambda$. We get

$$\hat{\psi}(\delta)=(g\otimes 1)\hat{\psi}(\delta)(g^{-1}\otimes 1)=\sum_j g\lambda_j(\delta)g^{-1}\otimes f_j$$

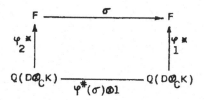

hence $\lambda_j(\delta)=g\lambda_j(\delta)g^{-1}$ hence $g\in Gal_\Lambda(D/C)$ and we are done.

(4.5) Theorem (4.4) above "reduces" $Gal_\Delta(F/K)$ to $Gal_\Lambda(D/C)$ which definitely is a more geometric object since Λ vanishes on C (while Δ does not vanish on K). But of course, in spite of its geometric nature the group $Gal_\Lambda(D/C)$ is still a misterious being in general, as well known from algebraic geometry (think of the Cremona groups and their subgroups defined by commutation with given rational vector fields on affine space).

(4.6) Notations being as in Theorem (4.4) put

$$Aut_\Delta(V/K)=\text{group of } \Delta\text{-automorphisms of } V/K$$

$$Aut_\Lambda(X/C)=\text{group of } \Lambda\text{-automorphisms of } X/C$$

Then clearly ψ^* induces a group isomorphism still denoted by

$$\psi^*:Aut_\Delta(V/K)\longrightarrow Aut_\Lambda(X/C)$$

But $Aut_\Lambda(X/C)$ is a closed subgroup of $Aut(X/C)$: $Aut_\Lambda(X/C)=\bigcap St(\theta)$ where θ runs through Λ and $St(\theta)$ is the stabilizer of θ in Aut(X/C) under the representation $Aut_{X/C}\longrightarrow GL(Der_C(\mathcal{O}_X))$, $\sigma\longmapsto\sigma^{**}$, $\sigma^{**}d=(\sigma^*)^{-1}d\sigma^*$. Note that the above representation is algebraic since its restriction to $G=Aut^o{}_{X/C}$ identifies via the isomorphism λ in (2.5) with the adjoit representation $Ad:G\longrightarrow GL(Lie(G))$. We conclude that $Aut_\Delta(V/K)$ identifies with the set of C-points of some group scheme locally of finite type over C. To illustrate this let's discuss the case $V=\mathbb{P}_K{}^n$. So suppose we are given a structure of Δ-variety on $V=\mathbb{P}_K{}^n=\mathbb{P}_C{}^n\otimes K$, $C=K^\Delta$. Under the identification $Der_K(\mathcal{O}_V)=sl(n+1,K)$ one immediately checks using (I.3.16) and Kolchin's expression for the logarithmic derivative of a linear algebraic group $[Kol_1]$p.352 that if $u:SL(n+1,K)\longrightarrow PGL(n,K)=Aut(V/K)$ is the natural projection then

$\Gamma := u^{-1}(\mathrm{Aut}_\Delta(V/K)) = \{g \in SL(n+1,K);\ (\delta g)g^{-1} + g\varphi(\delta)g^{-1} = \varphi(\delta)$ for all $\delta \in \Delta\}$

where $\varphi(\delta) = \delta_V - \delta_K^* \in sl(n+1,K)$. This group was called in [Cass] the stabilizer (of φ) under the Loewy action and clearly may be writen as

$$\Gamma = \{g \in SL(n+1,K);\ \delta g + [g,\varphi(\delta)] = 0 \quad \text{for all} \quad \delta \in \Delta\}$$

Note that in this form the group Γ appears as defined by linear differential equations of a pretty classical form. It was noted in [Cass] p. 947 that if K is universal in Kolchin's sense $[\mathrm{Kol}_1][\mathrm{Kol}_2]$ then the group Γ above is conjugated to a subgroup of $SL(n+1,C)$. A consequence of (4.4) is that this conjugation statement holds without the universality assumption (but K has still to be algebraically closed!). Since universal Δ-fields in differential algebra are quite artificial beings this consequence has some interest in itself. To prove it note that (4.4) implies that there exists $S \in SL(n+1,K)$ such that

$$S^{-1}\mathrm{Aut}_\Delta(V/K)S \subset PGL(n+1,C)$$

Since $\mathrm{Ker}(u) = \{1,-1\} \subset SL(n+1,C)$ we get $S^{-1}\Gamma S \subset SL(n+1,C)$ and we are done. Note however that one could replace (4.4) in the argument above by an ad-hoc specialisation argument based on (2.9).

(4.7) Now suppose in (4.4) and (4.6) that V is an abelian variety over K. In this particular case we have a pretty simple picture which we shall describe below. By [La] p.24 $\mathrm{Aut}(V/K)$ identifies with $\mathrm{Gal}(F/K)$ hence $\mathrm{Gal}_\Delta(F/K) \simeq \mathrm{Aut}_\Lambda(X/C)$. Now X is isomorphic to an abelian variety over C so if we fix some point $0 \in X(C)$ we may think of X as an abelian variety with zero emement 0. We denote by $\mathrm{End}(X)$ the endomorphism ring of the abelian variety $(X,0)$ and by $\mathrm{Aut}(X,0)$ the group of invertible elements of $\mathrm{End}(X)$; it is the set of all $a \in \mathrm{Aut}(X/C)$ for which $a(0)=0$. Finally put $\mathrm{Aut}_\Lambda(X,0) = \mathrm{Aut}(X,0) \cap \mathrm{Aut}_\Lambda(X/C)$. There is an obvious exact sequence

$$1 \longrightarrow Aut^{\circ}(X/C) \longrightarrow Aut_{\Lambda}(X/C) \longrightarrow Aut_{\Lambda}(X,0) \longrightarrow 1$$

so to understand $Gal_{\Delta}(F/K)$ one has to understand $Aut_{\Lambda}(X,0)$. Suppose in what follows that $C=\mathbb{C}$ and $X=\mathbb{C}^n/\Omega$ where Ω is some lattice in $\mathbb{C}^n$; we shall identify as usual $\mathbb{C}^n$ with $Lie(X)$ so Λ will be viewed as a subset of $\mathbb{C}^n$ and we have the following identifications

$$End(X)=\{B \in End(\mathbb{C}^n);\ B(\Omega)\subset\Omega\}$$

$$Aut_{\Lambda}(X,0)=\{B \in GL(\mathbb{C}^n);\ B(\Omega)\subset\Omega,\ Bx=x \text{ for all } x\in\Lambda\}$$

Then we claim that for any $B \in Aut_{\Lambda}(X,0)$ the following formula holds:

$$rank(B-1) \leqslant tr.deg.F^{\Delta}/K^{\Delta}$$

where 1 is the identity in $GL(\mathbb{C}^n)$. It is convenient to keep in mind that $tr.deg.F^{\Delta}/K^{\Delta}$ has a quite classical meaning; it is the maximum number of "independent algebraic prime integrals" cf.(IV.2), see also (3.22) above. If in the formula above we suppose $F^{\Delta}=K^{\Delta}$ then we get $rank(B-1)=0$ for all $B\in Aut_{\Lambda}(X,0)$ hence $Gal_{\Delta}(F/K)=Aut^{\circ}(X/C)$; this fact is well known and due to Kolchin [Kol_1]419; the above formula should be viewed as a generalisation of this fact. Let us prove our claim. Put $\Lambda(B)=Ker(B-1)$, $p=\dim_{\mathbb{C}}\Lambda(B)$. Clearly $\Lambda\subset\Lambda(B)$. On the other hand the vector space $\Lambda(B)$ is rational with respect to the lattice Ω (i.e. $\Lambda(B)$ has a $\mathbb{C}$-basis lying in Ω); this is clear since the matrix of B with respect to a $\mathbb{Z}$-basis of Ω has integer entries. Consequently $rank_{\mathbb{Z}}(\Omega\cap\Lambda(B))=\dim_{\mathbb{R}}\Lambda(B)$ so the torus $Y:=\Lambda(B)/\Lambda(B)\cap\Omega$ is an abelian variety in X. Consider the quotient map $u:X\longrightarrow X/Y$. We have that $u^*Q(X/Y)\otimes 1\subset Q(X\otimes K)^{\Delta}$ (where $X\otimes K$ is given a structure of Δ-variety via $\psi:V\longrightarrow X\otimes K$) as it will follow from the more general Lemma (4.8) below. Then we get

$$tr.deg.F^{\Delta}/\mathbb{C} \geqslant tr.deg.Q(X/Y)/\mathbb{C} = \dim X - \dim Y = n-p = rank(B-1)$$

and we are done.

Now we used (and shall use later too) the following easy remark:

(4.8) LEMMA. Let X be a projective variety over an algebraically closed field C, H an algebraic subgroup of $G=\mathrm{Aut}^{\circ}_{X/C}$ and $\mathrm{Lie}(H)$ its Lie algebra viewed as embedded in $\mathrm{Lie}(G)$. Suppose we have a dominant morphism of C-varieties $u:X_{\circ}\longrightarrow M$ where $X_{\circ}$ is an open subset of X such that $u^{-1}u(x)=X_{\circ}\cap Hx$ for all $x\in X_{\circ}(C)$. Then for any $f\in Q(M)$ and any $\theta\in\mathrm{Lie}(H)$ we have $(\lambda(\theta))(u^{*}f)=0$ where $\lambda:\mathrm{Lie}(G)\xrightarrow{\sim}\mathrm{Der}_C(\mathcal{O}_X)$ is the identification from (2.5).

Proof. One has to check that for sufficiently general $p\in X_{\circ}(C)$ the corresponding tangent vector $(\lambda(\theta))_p\in T_pX_{\circ}$ belongs to $T_p(u^{-1}u(p))=$ $=T_p(Hp)$. Consider the commutative diagram

$$\begin{array}{ccccc}
\mathrm{Lie}(G) & \xrightarrow{\sim} & T_eG & \xrightarrow{T_e\pi_p} & T_pX \\
\uparrow & & \uparrow & & \uparrow \\
\mathrm{Lie}(H) & \xrightarrow{\sim} & T_eH & \longrightarrow & T_p(Hp)
\end{array}$$

where e is the unit element in G, $T_e\pi_p$ is the tangent map of $\pi_p:G\longrightarrow X$, $\pi_p(g)=gp$ at e and $\mathrm{Lie}(G)\xrightarrow{\sim}T_eG$, $d\longmapsto d_e$ is the natural isomorphism. We have $(T_e\pi_p)(d_e)=\lambda(d)p$ for all $d\in\mathrm{Lie}(G)$ and this will close the proof. This equality follows immediately from the following commutative diagram ,see (2.5):

$$\begin{array}{ccccc}
\mathcal{O}_X & \longrightarrow & \mu_{*}\mathcal{O}_{G\times X} & \longrightarrow & \pi_{p*}\mathcal{O}_G \\
\lambda(d)\downarrow & & \downarrow d\otimes 1 & & \downarrow d \\
\mathcal{O}_X & \longrightarrow & \mu_{*}\mathcal{O}_{G\times X} & \longrightarrow & \pi_{p*}\mathcal{O}_G
\end{array}$$

(4.9) We close this § by examining the Lie algebra of $\mathrm{Aut}_{\Delta}(V/K)$. Suppose $G=\mathrm{Aut}^{\circ}_{X/C}$ in (4.4) and (4.6) is a linear algebraic group.

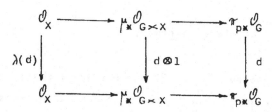

By [Che] p.172 we have

$$\mathrm{Lie}(\mathrm{Aut}_{\Lambda}(X/C)) = \Lambda^c := \{x \in \mathrm{Der}_C(\mathcal{O}_X);\ [x,y]=0 \text{ for all } y \in \Lambda\}$$

Hence $\mathrm{Aut}_{\Delta}(V/K)$ is discrete if and only if $\Lambda^c = 0$. Let us illustrate this remark by considering the case $V = \mathbb{P}^1$ (i.e. the Riccati equation). In this case $\mathrm{Lie}(G) = sl(2,C)$ and it is easy to check that if $\dim \Lambda \geq 2$ then $\Lambda^c = 0$. Hence if $\mathrm{Gal}_{\Delta}(F/K)$ is infinite then $\dim \Lambda \leq 1$. The case $\Lambda = 0$ leads to splitting. On the other hand there are only two 1-dimensional vector subspaces of $sl(2,C)$ up to equivalence under the adjoint representation and they correspond to the spaces $C\theta_o$ and $C\theta_1$ in $\mathrm{Lie}(G) \subset \mathrm{Der}_C C(t)$ where $\theta_o t = 1$ and $\theta_1 t = t$. Hence $\delta_F = a(\delta)\theta_i$ with $i \in \{1,2\}$ and $a(\delta) \in K$. So we find that F is generated over K by an element t such that

$$\text{either } \quad \delta t \in K \quad \text{ for all } \delta \in \Delta$$
$$\text{or } \quad t^{-1}\delta t \in K \quad \text{ for all } \delta \in \Delta$$

We obtained an alternative proof of a special case of Kolchin's theorem in [Kol_2] p.809 on the structure of Δ-function fields F/K of one variable for which $\mathrm{Gal}_{\Delta}(F/K)$ is infinite. Our proof may seem less direct but it certainly is more conceptual and somewhat shows what lies behind the computations in [Kol_2].

5. Descent of local complete Δ-rings.

Reminiscent of our method in (II.2) and (II.3) we shall solve here a problem left open in [Bu_2] namely the existence and differential-algebraic computation of the smallest algebraically closed field of definition for a local complete Noetherian ring. Same method leads to similar results for finitely dimensional Lie algebras.

Let K be a field. By a local complete ring over K we shall mean here a local Noetherian complete K-algebra A whose residue field A/m_A is a trivial extension of K. Such an A is always K-isomorphic to $K[[X_1,\dots,X_d]]/I$ where $d=\dim_K(m_A/m_A^2)$ and I is some ideal. Now a subfield C of K is called a field of definition for A if one can choose I above to be generated by elements of $C[[X_1,\dots,X_d]]$ (in other words if there exists a local complete ring A_o over C such that A is K-isomorphic to $A_o\hat{\otimes}_C K$, the completion of $A_o\otimes_C K$ with respect to $m_{A_o}\otimes_C K$; note that in this situation we have a flat morphism of local rings $A_o \longrightarrow A$ such that $m_{A_o}A=m_A$).

If K is a Δ-field then a local complete Δ-ring over K will mean a local complete ring A over K which is a Δ-ring and such that both natural morphisms $K \longrightarrow A$ and $A \longrightarrow K$ are Δ-morphisms (of course $A \longrightarrow K$ is a Δ-morphism if and only if m_A is a Δ-ideal).

Our main result will be:

(5.1) THEOREM. If K is an algebraically closed Δ-field and A is a local complete Δ-ring over K then K^{Δ} is a field of definition for A.

(5.2) COROLLARY. If K is an algebraically closed field and A is a local complete ring over K and if we put

$$\Delta(A)=\{\delta\in \mathrm{Der}(A);\ \delta(K)\subset K,\ \ \delta(m_A)\subset m_A\}$$

then the set of all algebraically closed fields of definition for A has a smallest element which equals $K^{\Delta(A)}$.

To prove Theorem (5.1) it is convenient to make first the following definition. Suppose K is a Δ-field and A is a local complete Δ-ring over K. We say that A is split (over K) if there exist a complete local ring A_o over $C:=K^{\Delta}$ and a K-isomorphism $\sigma : A \longrightarrow A_o \hat{\otimes}_C K$ such that $\sigma \delta_A \sigma^{-1}$ vanishes on $A_o = A_o \hat{\otimes} 1$ for all $\delta \in \Delta$. Such a σ will be called a splitting isomorphism.

(5.3) LEMMA. Let B be a split local complete Δ-ring over K and let $\sigma : B \longrightarrow B_o \hat{\otimes}_C K$ be a splitting isomorphism. Then $B^{\Delta} = \sigma^{-1}(B_o \hat{\otimes} 1)$. In particular B^{Δ} is a complete local ring over C and the natural morphism $B^{\Delta} \hat{\otimes}_C K \longrightarrow B$ is an isomorphism.

Proof. We may suppose σ is the identity. Clearly $B_o \hat{\otimes} 1 \subset B^{\Delta}$. To prove the converse inclusion take first any integer $n \geq 1$ and note that $B/m_B{}^n = B_o \otimes_C K/(m_{B_o} \otimes_C K)^n = (B_o/m_{B_o}{}^n) \otimes_C K$. We claim that we have $(B/m_B{}^n)^{\Delta} = B_o/m_{B_o}{}^n$. Indeed take a basis $e_1, \ldots, e_N$ of the C-vector space $B_o/m_{B_o}{}^n$ and if $b \in (B/m_B{}^n)^{\Delta}$, $b = \sum b_i e_i$ with $b_i \in K$ then for all δ, $0 = \delta b = \sum (\delta b_i) e_i$ hence all b_i belong to C. Now take $b \in B^{\Delta}$ and let's prove that $b \in B_o = B_o \hat{\otimes} 1$. Denote by $p_n : B \longrightarrow B/m_B{}^n$ the canonical projections. Since $\delta b = 0$ we get $0 = p_n \delta b = \delta p_n(b)$ hence $p_n(b) \in B_o/m_{B_o}{}^n$. So there exist elements $b_n \in B_o$ such that $b - b_n \in m_B{}^n$ hence $b_n \longrightarrow b$ in B. In particular the sequence $(b_n)_n$ is a Cauchy sequence in B. Since $m_B{}^n \cap B_o = m_{B_o}{}^n$ the sequence $(b_n)_n$ is also a Cauchy sequence in B_o. Since B_o is complete we have $b_n \longrightarrow b_o \in B_o$ hence $b = b_o \in B_o$ and we are done.

(5.4) PROPOSITION. Let K be a Δ-field with $C:=K^{\Delta}$ algebraically closed and let A be a local complete Δ-ring over K. Then there exists a sequence $K = K_1 \subset K_2 \subset K_3 \subset \ldots \subset F = \bigcup K_n$ of Δ-field extensions such that $F^{\Delta} = C$, K_n is finitely generated over K as a field extension and $A \hat{\otimes}_K F$ is split over F. Moreover if K is a partial Δ-field then one can take all K_n/K to be Picard-Vessiot extensions.

Proof. Step 1. Assume A is Artinian. In this case pick a K-basis $e_1, \ldots, e_N$ of m_A and write

$$\delta e_i = \sum_{j=1}^{N} a_{ij}(\delta)e_j \quad \text{for} \quad 1\leq i\leq N$$

View $a(\delta):=(a_{ij}(\delta))$ as a K-point of the Lie algebra $gl(N)$ of $G=GL(N)$. By (II.2.9) there exists a Δ-field extension F/K with F finitely generated over K as a field extension and with $F^{\Delta}=K^{\Delta}$ and there exists $g\in G(F)$ such that $\ell\delta g=a(\delta)$ for all $\delta\in\Delta$. Moreover if K is a partial Δ-field we may assume F/K is a Picard-Vessiot extension. By [Kol$_1$] p.352, $\ell\delta g=(\delta g)g^{-1}$ so we get $\delta g=a(\delta)g$. Let $f_i\in A\otimes_K F$ be defined by the equalities:

$$e_i = \sum_{k=1}^{N} g_{ik}f_k \qquad \text{for} \quad 1\leq i\leq N$$

We get

$$\sum_{jk} a_{ij}(\delta)g_{jk}f_k=\sum_{j} a_{ij}(\delta)e_j=\delta e_i=\sum_{k}(\delta g_{ik})f_k+\sum_{k} g_{ik}\delta f_k$$

Since $\delta g=a(\delta)g$ we get $\delta f_k=0$ for all k and δ. Now we may write

$$f_if_j=\sum_{k} c_{ijk}f_k, \qquad c_{ijk}\in F$$

Applying δ to the above equality we get

$$0=\delta(f_if_j)=\sum_{k}(\delta c_{ijk})f_k$$

for all i,j,δ hence $c_{ijk}\in C$. Put $A_o:=C+Cf_1+\dots Cf_N$. Then A_o is an Artinian local C-subalgebra of $A\otimes_K F$ with residue field C and the natural F-morphism $A_o\otimes_C F\longrightarrow A\otimes_K F$ is an isomorphism.

Step 2. Put $A_n=A/m_A{}^n$. By Step 1 there exist Δ-field extensions F_n/K finitely generated as field extensions and with $F_n^{\Delta}=K^{\Delta}$ such that $A_n\otimes_K F_n$ are split over F_n. Note that if K is a partial Δ-field and K^+ is a constrainedly closed extension of K with the same field of constants as K then one can choose all F_n to be subfields of K^+; in this case put $K_n=F_1F_2\dots F_n$ and note that K_n/K are Picard-Vessiot extensions. In the general case we may choose

$F_1\subset F_2\subset\dots$ and put $K_n=F_n$. In any case we have $K_1\subset K_2\subset\dots\subset F=\bigcup K_n$ and $A_n\otimes_K F$ are split over F. Put $A_{n0}=(A_n\otimes_K F)^{\Delta}$. By Lemma (5.3) the natural morphisms $\varphi_n: A_{n0}\otimes_C F\longrightarrow A_n\otimes_K F$ are isomorphisms. Since the natural surjections $A_{n+1}\otimes_K F\longrightarrow A_n\otimes_K F$ are Δ-morphisms they carry constants into constants and hence induce surjections $f_n: A_{n+1,0}\longrightarrow A_{n0}$. Clearly all Zariski tangent spaces to A_{n0} have dimension $d=\dim_K(m_A/m_A^{\,2})$. Choose a C-basis of the Zariski tangent space of A_{20} and lift it successively to all A_{n0}'s; this will give us a sequence of surjective maps $C[[X]]\xrightarrow{u_n}A_{n0}$ compatible with the maps f_n; here $X=(X_1,\dots,X_d)$. Let J_n be the kernel of u_n and put $J_0=\bigcap J_n$ $(1\leq n<\infty)$. If $A=K[[X]]/J$ then $A_n=K[[X]]/J+(X)^n$ hence $A_n\otimes_K F=(K[[X]]/J+(X)^n)\otimes_K F=F[[X]]/JF[[X]]+(X)^n$. Analogously $A_{n0}\otimes_C F=(C[[X]]/J_n)\otimes_C F=F[[X]]/J_nF[[X]]$ because J_n are (X)-primary. Now consider the diagram

$$\begin{array}{ccccccc}
\dfrac{F[[X]]}{\bigcap J_nF[[X]]} & \xrightarrow{\ i\ } & \varprojlim \dfrac{F[[X]]}{J_nF[[X]]} & \xrightarrow{\ \varphi\ } & \varprojlim \dfrac{F[[X]]}{JF[[X]]+(X)^n} & = & \dfrac{F[[X]]}{JF[[X]]} \\
 & & \downarrow{\scriptstyle p_2} & & \downarrow{\scriptstyle q_2} & & \\
 & & \dfrac{F[[X]]}{J_2F[[X]]} & \xrightarrow{\ \varphi_2\ } & \dfrac{F[[X]]}{JF[[X]]+(X)^2} & = & \dfrac{F[[X]]}{(X)^2}
\end{array}$$

where $\varphi=\varprojlim\varphi_n$. Clearly i is injective. On the other hand $p_2 i$ is the canonical projection hence $q_2\varphi i$ is surjective. Since $F[[X]]/JF[[X]]$ is (X)-adically complete this implies that φi itself is surjective so i is an isomorphism. We claim that $\bigcap J_nF[[X]]=J_0F[[X]]$, in other words that

$$\bigcap \frac{J_nF[[X]]}{J_0F[[X]]}=0 \qquad \text{in} \qquad S:=\frac{F[[X]]}{J_0F[[X]]}$$

Indeed $\bar{J}_n := J_n/J_o$ is a descending sequence of ideals in $R := C[[X]]/J_o$ whose intersection is zero. By [Nag] p.103 for any $n \geq 1$ there exists an integer $N(n)$ such that $\bar{J}_{N(n)} \subset m_R^{\,n}$. It follows that $\bar{J}_{N(n)} S \subset m_S^{\,n}$ hence $\bigcap \bar{J}_{N(n)} S \subset \bigcap m_S^{\,n} = 0$. But $\bar{J}_n S = J_n F[[X]]/J_o F[[X]]$ and our claim is proved. We conclude that we have an isomorphism $V = \varphi i : F[[X]]/J_o F[[X]] \longrightarrow F[[X]]/JF[[X]]$; moreover for each i we have $(V^{-1}\delta V)(X_i \bmod J_o F[[X]]) = 0$ hence $A \hat{\otimes}_K F$ splits over F.

(5.5) Proof of (5.1). Let $A = K[[X]]/J$, $X = (X_1, \ldots, X_d)$. By Proposition (5.4) there exists a Δ-field extension F/K with $F^{\Delta} = K^{\Delta}$ such that $F[[X]]/JF[[X]]$ is F-isomorphic to $F[[X]]/J_o F[[X]]$ for some $J_o \subset C[[X]]$, $C := K^{\Delta}$. A theorem of Seidenberg [Se$_3$] implies now that in fact $K[[X]]/J$ is K-isomorphic to $K[[X]]/J_o K[[X]]$ and we are done.

(5.6) One of the situations in which (5.1) applies is the following. Suppose K is an algebraically closed Δ-field and V is a Δ-variety over K (not necessarily projective !). Suppose $p \in V_{\Delta}$ is a closed point (for instance let $p \in V$ be an isolated singularity). Then K^{Δ} is a field of definition for $\mathcal{O}_{V,p}$ (although it is not in general a field of definition for any neighbourhood of p in V). Another significant situation in which (5.1) applies is when $K = (C((Z))_a, d/dZ)$, where the index "a" denotes as usual the algebraic closure. Let's look at the following special case of this situation. Let $f \in C[Z, X_1, \ldots, X_d]$ be a quasi-homogenous polynomial with respect to some integer weights $a, a_1, \ldots, a_d$ such that $f(Z,0) = 0$. By Euler's formula the derivation d/dZ of K lifts to a derivation

$$\delta = \partial/\partial Z + (a_1/a)X_1(\partial/\partial X_1) + \ldots + (a_d/a)X_d(\partial/\partial X_d)$$

of $A = K[[X_1, \ldots, X_d]]/(f)$ and clearly $\delta(m_A) \subset m_A$. On the other hand if we consider the K-automorphism σ of $K[[X_1, \ldots, X_d]]$ sending X_i into $Z^{a_i/a} X_i$ then $\sigma(f) = Z^{N/a} g$ where N is the total degree of f

with respect to the weights and $g \in C[X_1,\dots,X_d]$. Hence C is seen directly to be a field of definition for A. But this example , although trivial, shows that even if we start with a polynomial f as above and end up also with a polynomial as g it may be necessary to consider isomorphisms σ involving non-integral powers of Z i.e isomorphisms over $C((Z))_\bullet$ and not only over $C((Z))$.

We will close our discussion on local complete rings by looking at the descent of morphisms between them. In (5.7) and (5.8) we assume for simplicity that K is uncountable.

(5.7) COROLLARY. Let K be an algebraically closed Δ-field and $f:A \longrightarrow B$ a Δ-morphism of local complete Δ-rings over K. Then there exists a commutative diagram

$$\begin{array}{ccc} A & \xrightarrow{\quad f \quad} & B \\ \sigma \downarrow & & \downarrow \tau \\ A_\bullet \hat{\otimes}_C K & \xrightarrow{\;f_\bullet \hat{\otimes} 1\;} & B_\bullet \hat{\otimes}_C K \end{array}$$

where σ, τ are K-isomorphisms and $f_\bullet : A_\bullet \longrightarrow B_\bullet$ is a morphism of local complete rings over C.

Proof. Suppose $A=K[[X]]/I$, $B=K[[X]]/J$, $X=(X_1,\dots,X_p)$, $Y=(Y_1,\dots,Y_q)$. By Proposition (5.4) there exists a Δ-field extension F/K with $F^\Delta = K^\Delta = C$ and there exist splitting isomorphisms $\sigma_F : F[[X]]/IF[[X]] \longrightarrow F[[X]]/I_\bullet F[[X]]$ and $\tau_F : F[[Y]]/JF[[Y]] \longrightarrow F[[Y]]/J_\bullet F[[Y]]$ for some $I_\bullet \subset C[[X]]$ and $J_\bullet \subset C[[Y]]$. If we give $F[[X]]/I_\bullet F[[X]]$ and $F[[Y]]/J_\bullet F[[Y]]$ the obvious split structure of Δ-rings then σ_F and τ_F are Δ-isomorphisms hence $g:=\tau_F f \sigma_F^{-1}$ will be a Δ-morphism. In particular g takes constants into constants so by (5.3) g induces a morphism $f_\bullet : C[[X]]/I_\bullet \longrightarrow C[[Y]]/J_\bullet$ making the following diagram commutative:

$$(*_F)\qquad \begin{array}{ccc} F[[X]]/IF[[X]] & \xrightarrow{\;f\hat{\otimes}1_F\;} & F[[Y]]/JF[[Y]] \\ \sigma_F\big\downarrow & & \big\downarrow\tau_F \\ F[[X]]/I_{\circ}F[[X]] & \xrightarrow{\;f_{\circ}\hat{\otimes}1_F\;} & F[[Y]]/J_{\circ}F[[Y]] \end{array}$$

Now one can obviously write down a system S of polynomial equations with coefficients in K involving countably many unknowns such that that for any field extension L/K the following conditions are equivalent:

1) S has a solution in $L\times L\times\ldots$.

2) There exist L-isomorphisms σ_L, τ_L making the diagram $(*_L)$ commutative where $(*_L)$ is by definition the diagram obtained from $(*_F)$ by replacing F with L.

By what we just proved S has a solution in $F\times F\times\ldots$. Since K is algebraically closed and uncountable, S will have a solution in $K\times K\times\ldots$ by a result of Krull [Se_3]p.72 and we are done.

(5.8) COROLLARY. Let K be an algebraically closed Δ-field, $C:=K^{\Delta}$ and suppose we are given a structure of Δ-ring on $K[[X]]$, $X=(X_1,\ldots,X_d)$ extending that of K and such that $\delta(X)\subset(X)$ for $\delta\in\Delta$. Then there exists a K-automorphism σ of $K[[X]]$, a set of derivations $\Lambda\subset Der_C(C[[X]])$ with $\Lambda(X)\subset(X)$ and a group isomorphism

$$\sigma^{\#}:Aut_{\Delta}(K[[X]]/K)\longrightarrow Aut_{\Lambda}(C[[X]]/C)$$

such that for any $f\in Aut_{\Delta}(K[[X]]/K)$ the following diagram is commutative:

$$\begin{array}{ccc} K[[X]] & \xrightarrow{\;f\;} & K[[X]] \\ \sigma\big\downarrow & & \big\downarrow\sigma \\ K[[X]] & \xrightarrow{\;\sigma^{\#}(f)\hat{\otimes}1\;} & K[[X]] \end{array}$$

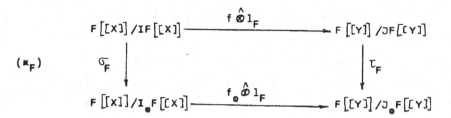

Remarks:1) In the above statement $\mathrm{Aut}_{\Delta}(K[[X]]/K)$ denotes of course the group of all K-automorphisms of $K[[X]]$ which commute with every member of Δ. Same meaning for $\mathrm{Aut}_{\Lambda}(C[[X]]/C)$.

2) $\mathrm{Aut}_{\Lambda}(C[[X]]/C)$ is more "geometric" than $\mathrm{Aut}_{\Delta}(K[[X]]/K)$ because Λ vanishes on C while Δ does not vanish on K; in particular $\mathrm{Aut}_{\Lambda}(C[[X]]/C)$ is described (as a subgroup of the group of all C-automorphisms of $C[[X]]$) by a system of "formal" partial algebraic differential equations with constant coefficients as we shall see below.

3) Even if we start with Δ consisting of one element only it may happen that Λ generates inside $\mathrm{Der}_C(C[[X]])$ an infinite dimensional Lie C-subalgebra !

Proof. Exactly as in the proof of (5.7) one can find $\sigma\in\mathrm{Aut}(K[[X]]/K)$ such that $\sigma^{-1}(\mathrm{Aut}_{\Delta}(K[[X]]/K))\sigma\subset\mathrm{Im}(\mathrm{Aut}(C[[X]]/C)\longrightarrow\mathrm{Aut}(K[[X]]/K))$. So we may suppose σ=identity. Write

$$\delta_{K[[X]]}=\delta_K^{*}+\sum a_{\delta j}(X)(\partial/\partial X_j)$$

where $a_{\delta j}(X)\in K[[X]]$ and δ_K^{*} is induced by the trivial lifting of δ_K to $C[[X]]\otimes_C K$. Clearly $a_{\delta j}(X)\in(X)$. Put $f_i=f(X_i)$ for any $f\in\mathrm{Aut}_{\Delta}(K[[X]]/K)$. Since $\delta f(X_i)=f(\delta X_i)$ we get

$$a_{\delta i}(f(X))=\sum a_{\delta j}(X)(\partial f_i/\partial X_j)$$

which can be viewed as a system of formal partial algebraic differential equations satisfied by $(f_1,\dots,f_d)$. Now for each δ the K-vector space generated by all coefficients of $a_{\delta 1},a_{\delta 2},\dots,a_{\delta d}$ is countably generated; let $e_{\delta 1},e_{\delta 2},\dots$ be a basis of it (which may be finite or at most countable). Then one can write $a_{\delta k}(X)=\sum e_{\delta p}a_{\delta kp}(X)$ with $a_{\delta kp}(X)\in C[[X]]$ and $a_{\delta kp}(X)\longrightarrow 0$ in $C[[X]]$. Since $f_i\in C[[X]]$ we get $a_{\delta ip}(f(X))=\sum a_{\delta jp}(X)(\partial f_i/\partial X_j)$ for all p. Then clearly the restriction of f to $C[[X]]$ belongs to $\mathrm{Aut}_{\Lambda}(C[[X]]/C)$ where

Λ is the Lie C-subalgebra of $\mathrm{Der}_C(C[[X]])$ generated by all derivations $D_{\delta p} := \sum a_{\delta j p}(X)(\partial/\partial X_j)$. Conversely if f belongs to the image of $\mathrm{Aut}_\Lambda(C[[X]]/C)$ in $\mathrm{Aut}(K[[X]]/K)$ then clearly $f \in \mathrm{Aut}_\Delta(K[[X]]/K)$ and we are done.

(5.9) Using the same method as in the proof of (5.1) one can treat the problem of descending and splitting finite dimensional Lie algebras. We give in what follows the definitions and main results without proofs. If K is a field, C is a subfield of K and L is a Lie K-algebra, we say that C is a field of definition for L if there exists a Lie C-algebra M such that L and $M \otimes_C K$ are isomorphic as Lie K-algebras. Now given a Lie K-algebra L we denote by Der(L) the set of all $\mathbb{Q}$-derivations of L (i.e. $\mathbb{Q}$-linear maps $d: L \longrightarrow L$ such that $d[x,y] = [dx,y] + [x,dy]$ for all $x,y \in L$). If $\delta \in \mathrm{Der}(K)$ and $d \in \mathrm{Der}(L)$ we say that d is compatible with δ if $d(ax) = (\delta a)x + a dx$ for all $a \in K$, $x \in L$.

If K is a Δ-field then by a Δ-Lie K-algebra we mean a Lie K-algebra L together with a map $\Delta \longrightarrow \mathrm{Der}(L)$, $\delta \longmapsto \delta_L$ such that for any $\delta \in \Delta$, δ_L is compatible with δ_K. Examples of Δ-Lie K-algebras are for instance $\mathrm{Lie}_K(G)$ and $\mathrm{Der}_C(\mathcal{O}_X) \otimes_C K$ from (II.2.6). A Δ-Lie K-algebra will be called split if there exists an isomorphism of Lie K-algebras $\sigma: L \longrightarrow M \otimes_C K$, $C := K^\Delta$, with M a Lie C-algebra, such that $\sigma\delta_L\sigma^{-1}$ vanishes on $M \otimes 1$ for all $\delta \in \Delta$. The examples from (II.2.6) are of course split.

(5.10) PROPOSITION. Let K be a Δ-field with K^Δ algebraically closed and L a finite dimensional Δ-Lie K-algebra. Then there exists a Δ-field extension F/K finitely generated as a field extension, such that $F^\Delta = K^\Delta$ and $L \otimes_K F$ is split over F. Moreover if K is a partial Δ-field then one can take F/K to be a Picard-Vessiot extension.

This answers a question from [NW] p.999 in the finite dimensional case.

(5.11) THEOREM. Let K be an algebraically closed Δ-field and L a finite dimensional Δ-Lie K-algebra. Then K^{Δ} is a field of definition for L.

(5.12) COROLLARY. Let K be an algebraically closed field and L a finite dimensional Lie K-algebra. Then the set of all algebraically closed fields of definition for L has a smallest element which equals $K^{\Delta(L)}$, where $\Delta(L)$ is the set of all $\delta \in Der(K)$ for which there exists at least a derivation in $Der(L)$ compatible with δ.

(5.13) We close this paragraph (and chapter !) by making the following remarks:

1) As proved in (II.2) and (II.5) various " Δ-objects over K" are split over a strongly normal extension of K, say F/K (or over an inductive limit of such extensions). Due to the fact that Δ-Galois theory of strongly normal extensions is well understood one can start looking for a " Δ-Galois cohomological" interpretation of F/K-forms of our Δ-objects as expected in [NW]p.999. We shall not pursue this idea here.

2) Unlike in [NW] our splitting results hold without assuming that our "Δ-objects" descend to constants ! On the contrary the moral of our tneory is that "Δ-objects" automatically descend to constants !

CHAPTER III. NORMALITY IN DIFFERENTIAL GALOIS THEORY.

In this chapter we prove our main results concerning the connections between the following 3 properties:

(WN) weak normality (see(I.3.1))

(SN) strong normality (see(I.3.1))

(NMS) no movable singularity (see(I.1.3)).

1. Reduction to same field of constants.

In normality questions one is dealing exclusively with the case of Δ-field extensions having the same field of constants. The aim ofthis § is to show that, as far as one is concerned with Δ-function fields with no movable singularity, one can always "reduce" the general case to the case above in the sense that we have:

(1.1) THEOREM. Let F/K be a Δ-function field with no movable singularity and with K algebraically closed. Let K_1 be the algebraic closure of $K(F^{\Delta})$ in F. Then F/K_1 has no movable singularity.

So we see that we have a decomposition:

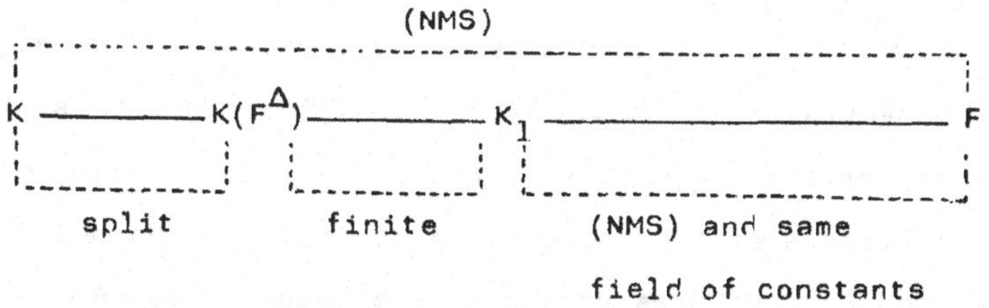

Since from our point of view split and finite extensions are in some sense trivial we see that Δ-function fields with (NMS) are"reduced" to those which in addition have the same field of constants. A further reduction is possible and useful, see (1.4).

To prove Theorem (1.1) we need the following easy algebraic result:

(1.2) LEMMA. Let A be a noetherian normal domain and $s_0,\ldots,s_n \in A$ non-zero elements such that the ideal $I=s_0A+\ldots+s_nA$ is proper and has height$(I)\geq 2$. Suppose we are given a derivation $\delta\in Der(A)$ such that $\delta(s_i/s_j)=0$ for all $0\leq i\leq j\leq n$. Then $\delta s_i\in s_iA$ for all $0\leq i\leq n$ in particular $\delta(I)\subset I$.

Proof. By normality of A we have for each i:

$$s_iA=\bigcap_P(s_iA_P\cap A)$$

the intersection being taken after all primes $P\in \mathrm{Spec}\,A$ of height 1 containing s_i. Take such a P; since $I\not\subset P$ there exists an index $j=$ $=j(P)$ such that $s_j\notin P$. Hence $s_iA_P=(s_i/s_j)A_P$ so $\delta(s_iA_P)\subset s_iA_P$. Consequently $\delta(s_iA)\subset s_iA$ and we are done.

(1.3) Now we start proving Theorem (1.1). By (II.1.24) there exists a non-singular projective Δ-model V of F/K. By [BS] p.94 there exists a non-singular projective model X of F^Δ/K^Δ. Choose a projective embedding $X\longrightarrow \mathbb{P}_C^N=\mathrm{Proj}\,C[T_n,\ldots,T_N]$, $C=K^\Delta$ and denote by $f:V\dashrightarrow X\otimes K\longrightarrow \mathbb{P}_K^N$ the natural composed rational map. Let U be the maximal open subset of V on which f is defined and $f_U:U\longrightarrow \mathbb{P}_K^N$ the induced morphism. Since V is smooth the complement $V\setminus U$ has codimension ≥ 2 in V, $f_U^*\mathcal{O}(1)\in \mathrm{Pic}(U)$ extends to some $L\in \mathrm{Pic}(V)$ and the sections $f_U^*T_i\in H^0(U,f_U^*\mathcal{O}(1))$ extend to sections $s_i\in H^0(V,L)$; here we view T_i as sections in $H^0(\mathbb{P}_K^N,\mathcal{O}(1))$. Let Z be the scheme of zeroes of the sections $s_0,\ldots,s_N$. By [Ha] p.168 if we blow up V along Z and if we denote this blowing up by $W\longrightarrow V$ then the composed rational map $W\longrightarrow V\dashrightarrow X\otimes K\longrightarrow \mathbb{P}_K^N$ is a morphism. We claim Z is a Δ-subscheme of V. Indeed choose Spec A

$\subset V$ an open affine set where the restriction of L is trivial and identify the restriction of L to Spec A with the structure sheaf of Spec A. Then $s_0,\ldots,s_N$ are identified with elements in A. Since we may suppose X is not contained in any hyperplane of $\mathbb{P}_C^N$ we may suppose no s_j is zero. Clearly Z is given in A by the ideal $I=s_0A+\ldots s_NA$. Now the morphism $U\cap \operatorname{Spec} A \longrightarrow X\otimes K \longrightarrow \mathbb{P}_K^N$ is defined by glueing the ring homomorphisms

$$K\left[T_0/T_j,\ldots,T_N/T_j\right] \xrightarrow{\ \pi\ } B_j\otimes K \xrightarrow{\ f^*\ } A\left[s_j^{-1}\right]$$

$$T_i/T_j \longmapsto \widehat{T_i/T_j}\otimes 1 \longmapsto s_i/s_j$$

where Spec $B_j=X\cap \operatorname{Spec} K\left[T_0/T_j,\ldots,T_N/T_j\right]$, π being the canonical surjection and f^* being a morphism of Δ-rings. Since $\delta(B\otimes 1)=0$ for all $\delta\in\Delta$ we get $\delta(s_i/s_j)=0$ for all i,j hence by Lemma (1.2) $\delta(I)\subset I$. By (I.2.6) W is a Δ-variety with respect to $\delta_{Q(V)}$, $\delta\in\Delta$. Let $W \longrightarrow W_1 \longrightarrow X\otimes K$ be the Stein factorisation of $W\longrightarrow X\otimes K$ [Ha] p.280. By (I.2.5) $K_1:=Q(W_1)$ is a Δ-subfield of $Q(W)=Q(V)=F$, W_1 is a Δ-model of K_1/K and $W\longrightarrow W_1$ is a Δ-morphism. Clearly K_1 is the algebraic closure of $Q(X\otimes K)=K(F^\Delta)$ in F. Finally F/K_1 has no movable singularity since it has a projective Δ-model:

$$W\times_{W_1}\operatorname{Spec} K_1$$

and the theorem is proved.

Now in (III.2) - (III.4) we shall consider in fact only Δ-function fields F/K for which $F^\Delta=K^\Delta$ and for which in addition either K^Δ or even K is algebraically closed. We may reduce ourselves to this situation via Theorem (1.1) and the following:

(1.4) Remark. Let F/K be a Δ-function field with $F^\Delta=K^\Delta$, K_1/K an algebraic extension with $(K_1)^\Delta$ algebraically closed and put $K_1F=F\otimes_K K_1$. Then $(K_1F)^\Delta=(K_1)^\Delta$. Moreover if F/K has no movable singularity then of course the same property will hold for K_1F/K_1.

The proof of this remark is immediate via (I.1.2). So we see that we have an embedding diagram

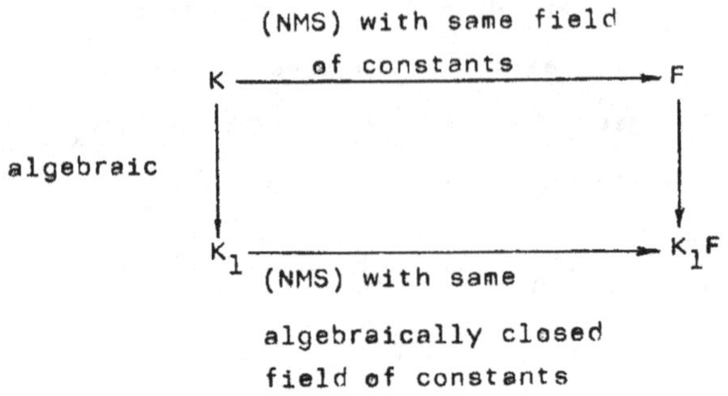

Two special cases of the above situation are useful in applications: first the case when K_1 is the algebraic closure K_a of K; the second is when $K_1 = K \otimes_{K^\Delta} (K^\Delta)_a$. In both cases we have $(K_1)^\Delta = (K^\Delta)_a = (K_a)^\Delta$.

2. Embedding (NMS) extensions into (SN) extensions.

The aim of this § is to prove and give applications of the following

(2.1) THEOREM. Let F/K be a partial Δ-function field such that K is algebraically closed and $F^\Delta = K^\Delta$. The following conditions are equivalent:

1) F/K has no movable singularity.

2) There exists a Picard-Vessiot extension E/F such that E/K is strongly normal.

Moreover if we are given a constrainedly closed extension F^+ of F with $(F^+)^\Delta = F^\Delta$ we may take E above to be contained in F^+.

To prove 1)$\Longrightarrow$2) in the above theorem we shall prove the following more precise embedding result (we shall need this more precise statement later):

(2.2) THEOREM. Let F/K be a partial Δ-function field with no movable singularity such that K is algebraically closed and $F^{\Delta}=K^{\Delta}$. Moreover let F^{+} be a constrainedly closed extension of F with $(F^{+})^{\Delta}=F^{\Delta}$ and let V be a projective Δ-model of F/K. Then there exist:

a) a Δ-subfield E of F^{+} containing F,

b) a descent isomorphism $\omega: V \longrightarrow X\otimes_C K$, $C=K^{\Delta}$,

c) a point $p\in X(C)$,

d) a connected algebraic subgroup G^{*} of $G=\mathrm{Aut}^{\circ}{}_{X/C}$ with $\dim G^{*}=$ $=\mathrm{tr.deg}.E/K$ and

e) a G^{*}-primitive $\beta\in G^{*}(E)$ over K with $E=K(\beta)$ such that the following diagram is commutative

$$\begin{array}{ccccc}
\mathrm{Spec}\ E & \xrightarrow{\quad\beta\quad} & G^{*}\hookrightarrow & & G \\
\downarrow & & & & \downarrow \pi_p \\
\mathrm{Spec}\ F \longrightarrow V & \xrightarrow{\ \omega\ } & X\otimes K & \longrightarrow & X
\end{array}$$

where $\pi_p(g)=gp$ and such that the isotropy group of p in G^{*} equals $\beta^{-1}G_{E/F}\beta$ (recall that by (I.3.18) we have an isomorphism $G_{E/K}\xrightarrow{\sim}G^{*}, \tau\mapsto\beta^{-1}\tau\beta$)

In particular $G^{*}p$ is Zariski open in X hence X appears as a $G_{E/K}$-equivariant completion of the homogenous space $G_{E/F}\backslash G_{E/K}$.

For terminology of equivariant completions see the statement of (2.4) below.

Proof. Take any descent isomorphism $\sigma: V\longrightarrow X\otimes K$ cf.(II.1.1). We claim that X has an open G-orbit. Indeed if this was not so we could find by (II.1.13) a dominant morphism of varieties $u: X_{\circ}\longrightarrow M$ with $X_{\circ}$ open in X such that $u^{-1}u(x)=(Gx)\cap X_{\circ}$ for all $x\in X_{\circ}(C)$ and such that $\dim M\geq 1$. By (II.4.8) we would get that $\mathrm{Der}_C(\mathcal{O}_X)$ vanishes on $u^*Q(M)$ hence that $(\sigma^*)^{-1}\delta_V\sigma^*$ vanishes on $(u^*Q(M))\otimes 1 \subset Q(X\otimes K)$ hence that $\sigma^*(u^*Q(M)\otimes 1)\subset F^{\Delta}$, contradicting $F^{\Delta}=C$. Choose any point $p\in X(C)$ lying in the open G-orbit of X and let $\pi_p:G\longrightarrow X$ be the morphism defined by $\pi_p(g)=gp$. This induces a dominant morphism $G\otimes K\longrightarrow X\otimes K\xrightarrow{\sigma^{-1}} V$. Now recalling from (II.2.6) that we have an isomorphism of Lie K-algebras $\lambda:\mathrm{Lie}_K(G)\longrightarrow \mathrm{Der}_C(\mathcal{O}_X)\otimes K$ and recalling from (II.1.22) the definition of $\hat{\sigma}:\Delta\longrightarrow \mathrm{Der}_C(\mathcal{O}_X)\otimes K$ let's introduce on $G\otimes K$ a structure of Δ-variety by putting

$$\delta_{G\otimes K}:=\delta_{KG}{}^* + \lambda^{-1}\hat{\sigma}(\delta)$$

where $\delta_{KG}{}^*$ is the trivial lifting of δ_K to $G\otimes K$. We claim that $G\otimes K\longrightarrow V$ is a morphism of Δ-varieties; indeed if we give $X\otimes K$ the structure of Δ-variety by putting

$$\delta_{X\otimes K}:=\delta_{KX}{}^* + \hat{\sigma}(\delta)$$

where $\delta_{KX}{}^*$ is the trivial lifting of δ_K to $X\otimes K$ then clearly $X\otimes K\longrightarrow V$ is a Δ-morphism while the fact that $G\otimes K\longrightarrow X\otimes K$ is a Δ-morphism follows from looking at the last diagram in the proof of (II.4.8). We also claim that $G\otimes K$ is a partial Δ-variety; indeed for all $\delta,\theta\in\Delta$ we have (cf.(II.2.6)):

$$\begin{aligned}[\delta_{G\otimes K},\theta_{G\otimes K}] &= [\delta_{KG}{}^*+\lambda^{-1}\hat{\sigma}(\delta),\theta_{KG}{}^*+\lambda^{-1}\hat{\sigma}(\theta)] = \\ &= \delta^{\#}(\lambda^{-1}\hat{\sigma}(\theta))-\theta^{\#}(\lambda^{-1}\hat{\sigma}(\delta))+[\lambda^{-1}\hat{\sigma}(\delta),\lambda^{-1}\hat{\sigma}(\theta)] = \\ &= \lambda^{-1}\delta^{\#}(\hat{\sigma}(\theta))-\lambda^{-1}\theta^{\#}(\hat{\sigma}(\delta))+\lambda^{-1}[\hat{\sigma}(\delta),\hat{\sigma}(\theta)] = \\ &= \lambda^{-1}[\delta_{X\otimes K},\theta_{X\otimes K}]=0\end{aligned}$$

Put in what follows $\varphi:\Delta\longrightarrow \mathrm{Lie}_K(G)$, $\varphi(\delta)=\lambda^{-1}\hat{\sigma}(\delta)$, $\delta\in\Delta$ and consider the F-scheme of finite type

$$S:=(G\otimes K)\times_V \mathrm{Spec}\ F$$

It is a partial Δ-scheme. By (I.3.10) there exists a commutative diagram of Δ-schemes

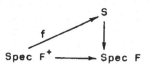

$$\begin{array}{ccc} & & S \\ & \overset{f}{\nearrow} & \downarrow \\ \operatorname{Spec} F^{+} & \longrightarrow & \operatorname{Spec} F \end{array}$$

By (I.2.4) the image q of $\operatorname{Spec} F^{+} \xrightarrow{f} S \longrightarrow G\otimes K$ belongs to $(G\otimes K)_{\Delta}$. Applying (II.2.8) to our G, q, φ we get that the residue field $K(q)$ of $G\otimes K$ at q (which equals the residue field of S at $f(\operatorname{Spec} F^{+})$) is a G-primitive extension of K with primitive given by $\operatorname{Spec} K(q) \longrightarrow G\otimes K \longrightarrow G$. Let E denote the image of $K(q)$ in F^{+} via f^{*}; then E/K is a G-primitive extension with G-primitive $\alpha:\operatorname{Spec} E \rightarrow \operatorname{Spec} K(q) \longrightarrow G\otimes K \longrightarrow G$ and moreover $F\subset E\subset F^{+}$. By (I.3.18) E/K is strongly normal and we have an injective homomorphism $c:G_{E/K} \longrightarrow G$ given by $c(\tau)=\alpha^{-1}\tau\alpha$ for $\tau\in G_{E/K}(C)$. Put $G^{*}= c(G_{E/K})$.

So far we produced an intermediate Δ-field E between F and F^{+}, a connected algebraic subgroup G^{*} of G of dimension $=\text{tr.deg.}E/K$ and a point $p\in X(C)$. Now we want to construct a G^{*}-primitive $\beta\in G^{*}(E)$ and a descent isomorphism $\omega:V \longrightarrow X\otimes K$ with the properties required in the Theorem. We proceed as follows.

Note that the equality $c(\tau)=\alpha^{-1}\tau\alpha$ reads also $\tau\alpha=\alpha c(\tau)=R_{c(\tau)}\alpha$ hence is expressed by the commutativity of the diagram

$$(*)\qquad \begin{array}{ccc} \operatorname{Spec} E & \xrightarrow{\ \alpha\ } & G\otimes K \\ {\scriptstyle \operatorname{Spec}\tau}\downarrow & & \downarrow{\scriptstyle R_{c(\tau)}} \\ \operatorname{Spec} E & \xrightarrow{\ \alpha\ } & G\otimes K \end{array}$$

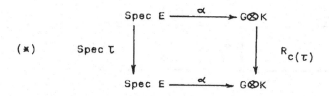

Let W be the closure of the image of $\alpha:\operatorname{Spec} E \longrightarrow G\otimes K$. Since $G\otimes K$ is a Δ-variety and α is a Δ-morphism, W has a natural structure of Δ-variety. By (*) W is stable under R_g for all $g\in G^*(C)$ hence for all $g\in G^*(K)$. Now we claim that the action $W\times G^*\longrightarrow W$ makes W a principal homogenous space. Indeed take $x\in W(K)$; we have $xG^*(K)\subset W(K)$ (multiplication being inside $G(K)$) and since we have the following estimate:

$$\dim(x(G^*\otimes K))=\dim(G^*\otimes K)=\dim G^* = \text{tr.deg.}E/K=\dim W$$

we get that $xG^*(K)=W(K)$ and our claim follows. Now it is trivial to check that we have a commutative diagram

$$\begin{array}{ccccccc}
\beta:\mathrm{Spec}\ E & \xrightarrow{\alpha} & W & \xrightarrow{L_x^{-1}} & G^*\otimes K & \longrightarrow & G^* \\
\downarrow & & \downarrow{\scriptstyle \pi'} & & \downarrow{\scriptstyle \pi\otimes 1} & & \\
\omega:\mathrm{Spec}\ F & \xrightarrow[\sigma']{} & X\otimes K & \xrightarrow[x^{-1}]{} & X\otimes K & &
\end{array}$$

where

$$\pi: G^* \hookrightarrow G \xrightarrow{\pi_p} X$$

$$\pi': W \hookrightarrow G\otimes K \xrightarrow{\pi_p\otimes 1} X\otimes K$$

$$\sigma': \mathrm{Spec}\ F \longrightarrow V \xrightarrow{\sigma} X\otimes K$$

At this point we constructed β and ω too. Clearly $E=K(\beta)$ and the diagram in the statement of the theorem is commutative. Let's show that β is a G^*-primitive over K. First give $G^*\otimes K$ the structure of Δ-variety induced via L_x^{-1} from that of W. Next remark that we have commutative diagrams

$$(**)\qquad \begin{array}{ccccc}
\mathrm{Spec}\ E & \xrightarrow{\alpha} & W & \xrightarrow{L_x^{-1}} & G^*\otimes K \\
{\scriptstyle \tau}\downarrow & & \downarrow{\scriptstyle R_{c(\tau)}} & & \downarrow{\scriptstyle R_{c(\tau)}} \\
\mathrm{Spec}\ E & \xrightarrow[\alpha]{} & W & \xrightarrow[L_x^{-1}]{} & G^*\otimes K
\end{array}$$

for all $\tau\in G_{E/K}(C)$. Now to see that β is a G^*-primitive over K it is sufficient by (II.2.8) to prove that $\delta_{G^*\otimes K}-\delta_K^*\in \mathrm{Lie}_K(G^*)$ hence that $\delta_{G^*\otimes K}$ commute with $R_{c(\tau)}{}^*$ for all $\tau \in G_{E/K}(C)$. But δ_E commute with all $\tau \in G_{E/K}(C)$ and we are done by (**). Note that (**) also implies that $c(\tau)=\beta^{-1}\tau\beta$ for all $\tau \in G_{E/K}(C)$.

To close the proof of our theorem let's show that the isotropy group H^* of p in G^* is $\beta^{-1}G_{E/F}\beta$ (or equivalently $c(G_{E/F})$). Let H be the isotropy group of p in G; then $H^*=H\cap G^*$. Now look at the diagram of fields deduced from (**):

$$\begin{array}{ccc} Q(G^*\otimes K) & \xrightarrow[\sim]{\beta^*} & E \\ {\scriptstyle (\pi\otimes 1)^*}\Big\uparrow & & \Big\uparrow \\ Q(X\otimes K) & \xrightarrow{\ \sim\ } & F \end{array}$$

Of course $Q(G^*)^{H^*(C)}=Q(X)$ with H^* acting on G^* via right translations and $Q(X)$ embedded in $Q(G^*)$ via π^*. By (I.3.2) we have $Q(G^*\otimes K)^{H^*(C)}=Q(X\otimes K)$. Now we are ready to prove that $H^*=c(G_{E/F})$. Indeed if $h\in H^*(C)$ then by the above equality R_h is the identity on $Q(X\otimes K)$ hence $\tau:=\beta^* R_h{}^*(\beta^*)^{-1}$ is the identity on F in other words $\tau\in G_{E/F}(C)$ hence $h=c(\tau)\in c(G_{E/F})$. Conversely if $\tau\in G_{E/F}(C)$ then $R_{c(\tau)}$ is the identity on $Q(G^*\otimes K)^{H^*(C)}$ hence on $Q(G^*)^{H^*(C)}$. By (I.3.6) $c(\tau)\in H^*(C)$ and the theorem is proved.

(2.3) Proof of 1)$\Longrightarrow$2) in Theorem (2.1). By (II.1.24) we may choose a non-singular projective Δ-model V of F/K. Apply Theorem (2.2) to this V and use notations from the statement of (2.2); clearly X will be non-singular. By results of Lieberman [Li$_2$] there is a natural homomorphism of algebraic groups

$$f:\mathrm{Aut}^\circ{}_{X/C} \longrightarrow \mathrm{Aut}^\circ{}_{\mathrm{Alb}(X)/C}$$

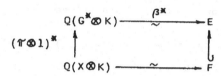

whose kernel is linear (in [Li$_2$] the analytic category was considered but one easily checks that the above statement holds in the algebraic category as well). Now if p is as in (2.2) let $\alpha_p:X\longrightarrow \mathrm{Alb}(X)$ be the Albanese mapping sending p into zero and let $\pi_p:\mathrm{Aut}^\circ{}_{X/C}\longrightarrow X$ be given by $\pi_p(g)=gp$ as in (2.2). It is a general fact that there is a commutative diagram

$$\begin{array}{ccc} \mathrm{Aut}^{\circ}_{X/C} & \xrightarrow{\ f\ } & \mathrm{Aut}^{\circ}_{\mathrm{Alb}(X)/C} \\ \downarrow{\scriptstyle \pi_p} & & \downarrow{\scriptstyle \wr} \\ X & \xrightarrow[\ \alpha_p\ ]{} & \mathrm{Alb}(X) \end{array}$$

By (2.2) $\alpha_p \pi_p(\beta^{-1} G_{E/F} \beta)=0$ consequently $\beta^{-1} G_{E/F} \beta \subset \mathrm{Ker}(f)$ hence $G_{E/F}$ is linear and implication $1) \Longrightarrow 2)$ is proved.

To prove the converse implication in Theorem (2.1) we shall prove first a result on equivariant completions:

(2.4) PROPOSITION. Let G be a connected algebraic group over an algebraically closed field C and $H \subset G$ a linear algebraic subgroup. Then the homogeneous space $H\backslash G$ (of left cosets xH, $x \in G$) admits a projective G-equivariant completion i.e. there exist a projective variety X over C, an action $G \times X \longrightarrow X$ and a G-equivariant open immersion $H\backslash G \longrightarrow X$.

Remark. The statement above was previously known at least in the following 2 cases:

1) when G is linear (Chevalley's theorem [Hu] p.80) and

2) when G is commutative (Serre [S_1]).

Proof. Let L be the maximal connected linear subgroup of G and put $B=LH$. We claim that B is linear, normal and $A:=G/B$ is an abelian variety. Indeed $LH=\mathrm{Ker}(G \xrightarrow{\ p\ } G/L \longrightarrow (G/L)/p(H))$ so LH is normal; $G/LH \simeq (G/L)/p(H)$ so G/LH is an abelian variety. To see that LH is linear it is sufficient to see by [Ro] p.430 that $(LH)^{o}$ is linear, where the upper "o" denotes the connected component of the identity. Now by [Ro] p.420 $LH/L \simeq H/L \cap H$ and by [Ro] p.439 $H^{o} \subset L$ so $H/L \cap H$ is finite hence $[LH:L] < \infty$. Since L is connected $(LH)^{o}=L$

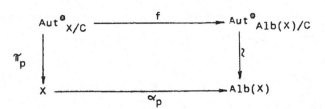

and our claim is proved. Now by Chevalley's theorem [Hu] p.80 there exists a rational representation $\rho: B \longrightarrow GL(E)$ where E is some finite dimensional C-vector space such that if we consider the associated projective action $\sigma: B \times P \longrightarrow P := \mathbb{P}(E)$, there exists $p_0 \in P(C)$ such that the isotropy group of p_0 in B is H. Write $\sigma(b,p) = bp$ for all $b \in B$ and $p \in P$. Define now two actions

$$\tau : B \times (G \times P) \longrightarrow G \times P$$

$$\theta : G \times (G \times P) \longrightarrow G \times P$$

by the formulae $\tau(b,(g,p)) = (gb^{-1}, bp)$ and $\theta(x,(g,p)) = (xg,p)$. Clearly these two actions commute. Consider also the action

$$\eta : B \times G \longrightarrow G$$

given by $\eta(b,g) = gb^{-1}$; clearly the projection map $\pi: G \longrightarrow A$ is a principal fibre bundle for η i.e. the natural map $G \times B \longrightarrow G \times_A G$ given by $(g,b) \longmapsto (gb^{-1}, g)$ is an isomorphism. Moreover the projection $\pi_1 : G \times P \longrightarrow G$ is G-equivariant for the actions τ and η. We claim there is a cartesian diagram of schemes

$$\begin{array}{ccc} G \times P & \xrightarrow{\ u\ } & Z \\ {\scriptstyle \pi_1}\downarrow & & \downarrow{\scriptstyle v} \\ G & \xrightarrow[\ \pi\ ]{} & A \end{array}$$

with Z quasi-projective over C and u a principal fibre bundle with group B for the action τ. By [Mum] pp.127-128 this will hold if we can find an invertible sheaf $\mathcal{L} \in \mathrm{Pic}(G \times P)$ ample relative to π_1 and a B-linearisation of $\mathcal{L}$ (with respect to τ). Take $\mathcal{L} = \pi_2^{*}\mathcal{O}_P(1)$ where $\pi_2 : G \times P \longrightarrow P$ is the second projection. It is sufficient to lift the action τ to an action of B on the line bundle $V(\mathcal{L}^{-1})$ associated to $\mathcal{L}^{-1}$ compatible with the projection $V(\mathcal{L}^{-1}) \longrightarrow G \times P$.

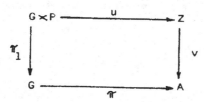

But $V(\mathcal{L}^{-1})$ naturally embeds into $G\times P\times E$ and B acts on $G\times P\times E$ by $(b,(g,p,e))\longmapsto(gb^{-1},bp,\rho(b)e)$ letting $V(\mathcal{L}^{-1})$ globally invariant. Our claim is proved. Now since π_1 is proper and smooth so will be v [Mi] p.20 hence Z is non-singular and projective. Using commutativity of τ and θ we check that θ descends to an action $\bar{\theta}$

$$\begin{array}{ccc} G\times(G\times P) & \xrightarrow{\ \theta\ } & G\times P \\ {\scriptstyle 1\times u}\downarrow & & \downarrow{\scriptstyle u} \\ G\times Z & \xrightarrow[\ \bar{\theta}\ ]{} & Z \end{array}$$

Let $x_o=u(1,p_o)\in Z$. It is trivial to check that the isotropy group of x_o in G with respect to $\bar{\theta}$ is H. Then we may conclude by letting X to be the closure of the orbit Gx_o of x_o in Z.

(2.5) Proof of 2) $\Longrightarrow$ 1) in Theorem (2.1). By Kolchin's theory (I.3.19) E/K is a full $G_{E/K}$-extension with $G_{E/K}$-primitive $\alpha\in G_{E/K}(E)$ By (I.3.17) and (I.3.18) α induces an isomorphism $\alpha^*:Q(G_{E/K}\otimes K)\longrightarrow E$ and the action of $Gal_\Delta(E/K)$ induces via α^* precisely the action of $G_{E/K}(C)$ by right translations on $Q(G_{E/K}\otimes K)$. By Proposition (2.4) there exists a $G_{E/K}$-equivariant completion X of $G_{E/F}\backslash G_{E/K}$. By (I.3.4) and (I.3.2) we get

$$\alpha^*F=\alpha^*(E^{Gal_\Delta(E/F)})=Q(G_{E/K}\otimes K)^{G_{E/F}(C)}=Q(X\otimes K)$$

By (I.3.17) We have $(\alpha^*)^{-1}\delta_E\alpha^*=\delta_K{}^*+\ell\delta\alpha$; let's introduce on $G_{E/K}\otimes K$ a structure of Δ-variety by putting $\delta_{G_{E/K}\otimes K}=\delta_K{}^*+\ell\delta\alpha$. With this structure α is a Δ-isomorphism. We claim that the derivation $\ell\delta\alpha$ takes $\mathcal{O}_{X\otimes K}$ into itself and this will close the proof. Of course it is sufficient to check that $Lie_C(G_{E/K})$ takes $\mathcal{O}_X$ into itself; but this follows from a diagram similar to the one at the end of the proof of (II.4.8). Theorem (2.1) is proved.

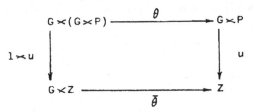

Now we discuss some consequences of Theorems (2.1) and (2.2).

(2.6) COROLLARY. Let L/K be a partial Δ-field extension with K algebraically closed and $K^{\Delta}=L^{\Delta}$ and let F_1, F_2 be two intermediate Δ-fields between K and L.

a) If both F_1/K and F_2/K have no movable singularity then the same holds for F_1F_2/K.

b) If $F_1 \subset F_2$, $[F_2:F_1] < \infty$ and F_2/K has no movable singularity then F_1/K still has no movable singularity.

Proof. a) By (I.3.11) there exists a constrainedly closed extension L^+ of L with $(L^+)^{\Delta}=L^{\Delta}$. By Theorem (2.1) there exist Δ-subfields E_1, E_2 in L^+ such that $E_i \supset F_i$ for $i=1,2$, E_i/F_i are Picard-Vessiot extensions and E_i/K are strongly normal. We have a diagram of Δ-subfields of L^+:

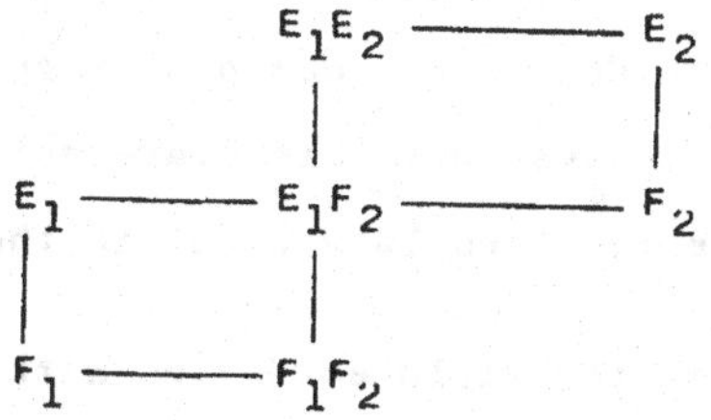

By (I.3) E_1E_2/K is strongly normal, E_1E_2/E_1F_2 is strongly normal with $G_{E_1E_2/E_1F_2} \simeq G_{E_2/E_2 \cap E_1F_2} \subset G_{E_2/F_2}$, E_1F_2/F_1F_2 is strongly normal with $G_{E_1F_2/F_1F_2} \simeq G_{E_1/E_1 \cap F_1F_2} \subset G_{E_1/F_1}$ and we have an exact sequence:

$$1 \longrightarrow G_{E_1E_2/E_1F_2} \longrightarrow G_{E_1E_2/F_1F_2} \longrightarrow G_{E_1F_2/F_1F_2} \longrightarrow 1$$

Since the first and the third groups are linear it follows by [Ro] p. 440 that the middle group is also linear and we are done by applying Theorem (2.1) again.

b) By Theorem (2.1) there is a Picard-Vessiot extension E/F_2 such that E/K is strongly normal. Since G_{E/F_2} has finite index in G_{E/F_1} it follows that G_{E/F_1} is still linear and we conclude by (2.1) again.

Remark. The above result was proved in [Mtd]p.101 in the case cardΔ =tr.deg.L/K=1 by a different method (which was intimately related to the "curve case"). However in [Mtd] the condition $K^{\Delta}=L^{\Delta}$ was not assumed.

(2.7) COROLLARY. Let L/K be a partial Δ-function field with K algebraically closed and $L^{\Delta}=K^{\Delta}$. Then the set of intermediate Δ-fields F between K and L such that F/K has no movable singularity has a largest element.

Proof. It is an immediate consequence of (2.6) since the lattice of intermediate fields between K and L is noetherian.

Remark. The case card Δ=tr.deg.L/K=1 was proved in [Mtd]p.101 without the assumption $L^{\Delta}=K^{\Delta}$; note that this statement was proved in [Mt first and then used to deduce our Corollary (2.6) in the "curve case".

Let F/K be a finitely generated field extension with K algebraically closed. Denote by q(F/K) the irregularity of F/K i.e. the number $q(V):=\dim_K H^1(V,\mathcal{O}_V)$ where V is any non-singular projective model of F/K. It is a birational invariant, see [U]. Moreover denote by Alb(F/K) the subfield $a^*Q(\mathrm{Im}(a:V\longrightarrow \mathrm{Alb}(V)))$ of F which is also a birational invariant; clearly tr.deg.Alb(F/K)/K=α(F/K), see (II.3.14). Finally recall that F/K is called ruled if F is a purely transcendental extension of some intermediate field between K and F.

(2.8) COROLLARY. Let F/K be a partial Δ-function field with K algebraically closed and $F^{\Delta}=K^{\Delta}$. The following are equivalent:

1) F/K has no movable singularity and q(F/K)=0.

2) There exists a partial Δ-field extension E/F such that E/K is a Picard-Vessiot extension.

Moreover if we are given a constrainedly closed extension F^{+} of F with $(F^{+})^{\Delta}=F^{\Delta}$ then one can take E above to be contained in F^{+}.

Proof. 2)$\Longrightarrow$1) is clear from Theorem (2.1) plus the general inequality $q(F/K)\leqslant q(E/K)$. To prove 1)$\Longrightarrow$2) apply Theorem (2.2) plus the fact that if X is a non-singular projective variety over C with $q(X)=0$ then $\mathrm{Aut}^{\circ}_{X/C}$ is a linear algebraic group; this fact is of course a consequence of Lieberman's result $[Li_2]$ that the kernel of $\mathrm{Aut}^{\circ}_{X/C} \longrightarrow \mathrm{Alb}(X)$ is linear.

(2.9) COROLLARY. Let F/K be a partial Δ-function field with K algebraically closed and $F^{\Delta}=K^{\Delta}$. The following are equivalent:

1) F/K has no movable singularity and is not ruled.

2) F/K is an abelian extension.

Proof. 2)$\Longrightarrow$1) follows from Theorem (2.1). To prove 1)$\Longrightarrow$2) we apply Theorem (2.2); by a theorem of Matsusaka-Mumford [MM] if V is not ruled, $\mathrm{Aut}^{\circ}_{V/K}$ is an abelian variety hence so is $\mathrm{Aut}^{\circ}_{X/C}$ hence so is G^{*}, notations being as in (2.2). Since G^{*} is commutative and acts faithfully on X it follows that $G_{E/F}=1$ hence $F=E$ and we are done.

(2.10) Remark. Theorem (2.2) shows that if V is a non-singular projective Δ-variety over an algebraically closed Δ-field K and if $Q(V)^{\Delta}=K^{\Delta}$ then $V\simeq X\otimes_{C}K$, $C=K^{\Delta}$ with X a quasi-homogenous variety over C for the group $G_{E/K}$ (a non-singular projective variety is called quasi-homogenous for an algebraic group acting on it if it contains an open orbit cf. $[Li_2]$). On the other hand the structure of quasi-homogenous varieties is well understood through papers like [BO], $[Li_2]$ a.s.o. In particular it follows from these papers that for such an X the Albanese map $X \longrightarrow Alb(X)$ is surjective and its fibres are connected non-singular varieties with irregularity zero. Let us use this remark to prove the following:

(2.11) COROLLARY. Suppose the equivalent conditions from Theorem (2.1) hold. Then $Alb(F/K)$ is a Δ-subfield of F, it is algebraically closed in F and $\alpha(F/K)=q(F/K)$. Moreover the algebraic closure of $Alb(F/K)$ in E equals the maximal abelian extension of K contained in E.

Proof. By the remark above there is a surjective morphism $V \longrightarrow Alb(V)$ with $f_*\mathcal{O}_V=\mathcal{O}_{Alb(V)}$ and $q(V\times_{Alb(V)}\mathrm{Spec}A_a)=0$ where V is any non-singular projective Δ-model of F/K (cf. (II.1.24)) and A_a is an algebraic closure of $A:=Alb(F/K)$. By (I.2.5) A is a Δ-subfield of F and $Alb(V)$ is a Δ-model of A/K. So by (2.9) A/K is an abelian extension. Clearly A is algebraically closed in F and $\alpha(F/K)=q(F/K)$. Note also that if $A_aF=F\otimes_A A_a$ then A_aF/A_a has no movable singularity (a projective Δ-model being $V\times_{Alb(V)}\mathrm{Spec}\,A_a$) and $q(A_aF/A_a)=0$. Take (cf.(I.3.11)) any constrainedly closed extension E^+ of E with $(E^+)^{\Delta}=E^{\Delta}$. Since E^+ is in particular algebraically closed, A_aF embeds over F into E^+ and now by (2.8) we get a diagram of Δ-fields

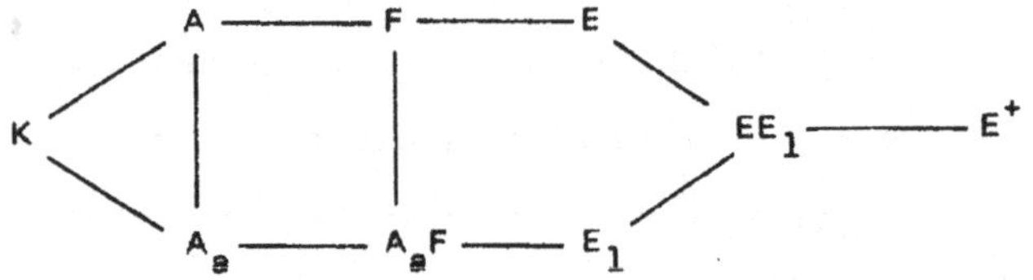

with E_1/A_a a Picard-Vessiot extension. Now EE_1/E_1 is strongly normal with $G_{EE_1/E_1}=G_{E/E\cap E_1}\subset G_{E/F}$ and A_aE/A_a is strongly normal with $G_{A_aE/A_a}=G_{E/E\cap A_a}$ hence EE_1/A_a is strongly normal and its group is still linear by [Ro]p.440. In particular $G_{E/E\cap A_a}$ is linear and connected. On the other hand since $[E\cap A_a : A] < \infty$, $E\cap A_a/K$ is an abelian extension. Consequently $E\cap A_a = E^L$ where L is the maximal connected linear subgroup of $G_{E/K}$; as one immediately checks E^L is in its turn precisely the maximal abelian extension of K contained in E and we are done.

3. (SN) is equivalent to (WN)+(NMS).

The aim of this § is to prove the following characterisation of strong normality:

(3.1) THEOREM. Let F/K be a Δ-function field. The following conditions are equivalent:

1) F/K is strongly normal.

2) F/K is weakly normal and has no movable singularity.

(3.2) Remarks. 1) One cannot remove the assumption of F/K having (NMS) in condition 2) above because of Kolchin's counterexample (I. 3.5). The counterexample has tr.deg.F/K=2 and $\varkappa(F/K)=-\infty$; we shall prove in (III.4) that the condition (NMS) can be in fact removed if tr.deg.F/K=1 or if tr.deg.F/K=2 and $\varkappa(F/K)\neq -\infty$.

2) One cannot remove the assumption of F/K being (WN) in condition 2) above; indeed take any Picard-Vessiot extension E/K and any closed subgroup $H\subset G_{E/K}$ which is not normal. Then E^H/K has no movable singularity by (2.1) and it is not strongly normal; in addition $(E^H)^\Delta = K^\Delta$.

3) The above theorem is an immediate consequence of (2.1) and (I. 3.4) provided K is algebraically closed. So what we really must solve is a Galois descent problem.

(3.3) It is convenient to start by recalling some basic constructions from [S2]. Let G be a finite group and A a G-group; write the action $G\times A\longrightarrow A$ in the form $(s,x)\longmapsto {}^sx$ (hence ${}^s(xy)={}^sx{}^sy$). Put as usual $Z^1(G,A)=\{\rho:G\longrightarrow A;\ \rho(st)=\rho(s)^s(\rho(t))\}$ and call the elements of $Z^1(G,A)$ cocycles. For $\rho,\rho'\in Z^1(G,A)$ write $\rho\sim\rho'$ if there exists $b\in A$ such that $\rho'(s)=b^{-1}\rho(s)b$ for all $s\in G$. Finally put $H^1(G,A)=Z^1(G,A)/\sim$.

Now let K be a field and K'/K a finite Galois extension. If W is a variety over K (recall that all our varieties are assumed quasi-projective cf. (I.1.1)) then denote by $\mathrm{Forms}(K'/K,W)$ the set of K-isomorphism classes of varieties V over K such that $V\otimes_K K'$ and $W\otimes_K K'$ are K'-isomorphic; elements of $\mathrm{Forms}(K'/K,W)$ are called K'/K-forms of W. It is well known that there is a natural bijection

$$\mathrm{Forms}(K'/K,W)\simeq H^1(\mathrm{Gal}(K'/K),\mathrm{Aut}(W\otimes_K K'/K'))$$

Let us however recall briefly the construction of this bijection since we shall need its explicit form. If $\rho:\mathrm{Gal}(K'/K)\longrightarrow \mathrm{Aut}(W\otimes K'/K')$ is a cocycle then define a group homomorphism $\hat{\rho}:\mathrm{Gal}(K'/K)\longrightarrow \mathrm{Aut}(W\otimes K'/K)$ by taking for each $s\in \mathrm{Gal}(K'/K)$:

$$\hat{\rho}(s):W\otimes K'\xrightarrow{1\otimes s} W\otimes K'\xrightarrow{\rho(s)} W\otimes K'$$

and define ${}_{\rho}W=W\otimes K'/\hat{\rho}$. By [Kn] p.180 W is still quasi-projective. Conversely if V is a representative of an element in $\mathrm{Forms}(K'/K,W)$ choose any isomorphism $\alpha:W\otimes K'\longrightarrow V\otimes K'$ over K' and define a cocycle ρ by the formula $\rho(s)(1\otimes s)=\alpha^{-1}(1\otimes s)\alpha$, $s\in \mathrm{Gal}(K'/K)$.

Now given a connected algebraic group A over K let $\mathrm{PHS}(K'/K,A)$ denote the set of K-isomorphism classes of principal homogeneous spaces W over K for A/K for which $W(K')\neq\phi$. It is well known that there is a natural bijection

$$\mathrm{PHS}(K'/K,A)\simeq H^1(\mathrm{Gal}(K'/K),A(K'))$$

but let's say a few words about this identification. The arrow from H^1 to PHS is constructed as follows: if ρ is a cocycle we consider the group homomorphism $\hat{\rho}:\mathrm{Gal}(K'/K) \longrightarrow \mathrm{Aut}(A\otimes K'/K)$, $\hat{\rho}(s)=L_{\rho(s)}\circ(1\otimes s)$ and put $W={}_{\rho}A:=A/\hat{\rho}$. The arrow from PHS to H^1 is constructed as follows: given a PHS W/K choose any point in $W(K')$ and take the multiplication morphism $A\otimes K' \longrightarrow W\otimes K'$ which is a K'-isomorphism; the cocycle corresponding to this isomorphism takes values actually in $A(K')$ (and not only in $\mathrm{Aut}(A\otimes K'/K')$!)its class being the image of W in H^1.

(3.4) To formulate our next result (which is the key ingredient in the proof of Theorem (3.1)) we consider the following setting. Let C be an algebraically closed field, K a field extension of C, K'/K a finite Galois extension, G a connected algebraic group over C, X a projective variety over C, $G\times X \longrightarrow X$ an action of G on X, and $x_o\in X(C)$ a point. Suppose that x_o has trivial isotropy group and the orbit Gx_o is open in X; so the map $\pi:G \longrightarrow X$, $\pi(g)=gx_o$ induces a C-isomorphism $\pi^*:Q(X) \longrightarrow Q(G)$. For any point $h\in G(C)$ right translation R_h induces an automorphism $\pi^{**}R_h^*\in \mathrm{Gal}(Q(X)/C)$, $\pi^{**}R_h^*:=(\pi^*)^{-1}R_h^*\pi^*$. We shall denote also by $\pi^{**}R_h^*$ the induced automorphism in $\mathrm{Gal}(Q(X\otimes K')/K')$. Finally for any cocycle $\rho\in Z^1(\mathrm{Gal}(K'/K),\mathrm{Aut}(X\otimes_C K'/K'))$ put as usual ${}_{\rho}(X\otimes K):=X\otimes K'/\hat{\rho}$ and view $Q({}_{\rho}(X\otimes K))$ as a subfield of $Q(X\otimes K')$.

(3.5) PROPOSITION. In notations above suppose there exists a Zariski dense subset S of $G(C)$ such that for all $g\in S$ we have

$$(\pi^{**}R_g^*)(Q({}_{\rho}(X\otimes K)))=Q({}_{\rho}(X\otimes K))$$

where $\pi^{**}R_g^*$ are viewed as elements in $\mathrm{Gal}(Q(X\otimes K'/K')$. Then ρ belongs to the image of the natural map

$$Z^1(\mathrm{Gal}(K'/K),G(K')) \longrightarrow Z^1(\mathrm{Gal}(K'/K),\mathrm{Aut}(X\otimes_C K'/K'))$$

Proof. Let Γ be the image of $\hat{\rho}:Gal(K'/K) \longrightarrow Aut(X\otimes K'/K)$ and Γ^* the image of Γ in $Gal(Q(X\otimes K')/K)$ via the natural map $\gamma \longmapsto \gamma^*$. We have $Q(\rho(X\otimes K)) = Q(X\otimes K')^{\Gamma^*}$. For all $g\in S$, all $x\in Q(\rho(X\otimes K))$ and all $\gamma\in\Gamma$ we have $\gamma^*(\pi^{**}R_g^*)x=(\pi^{**}R_g^*)x$ hence by Galois theory of finite extensions $(\pi^{**}R_g^*)^{-1}\gamma^*(\pi^{**}R_g^*)\in\Gamma^*$. So for all $\sigma\in Gal(K'/K)$ and all $g\in S$ there exists $\tau\in Gal(K'/K)$ depending on σ and g such that

$$(1)\qquad (\pi^{**}R_g^*)^{-1}\hat{\rho}(\sigma)^*(\pi^{**}R_g^*)=\hat{\rho}(\tau)^*$$

in $Gal(Q(X\otimes K')/K)$. We claim that the formula above can hold only if $\sigma=\tau$. Indeed since $\rho(\sigma)^*$ and $\pi^{**}R_g^*$ are the identity on $K'=1\otimes K'$ equality (1) yelds for any $z\in K'$

$$\sigma z=(\pi^{**}R_g^*)^{-1}(1\otimes\sigma)\rho(\sigma)^*(\pi^{**}R_g^*)z=$$

$$=(1\otimes\tau)\rho(\tau)^*z=\tau z$$

hence $\sigma=\tau$. Now since $1\otimes\sigma$ and $\pi^{**}R_g^*$ commute, (1) yelds

$$(2)\qquad (\pi^{**}R_g^*)^{-1}\rho(\sigma)^*(\pi^{**}R_g^*)=\rho(\sigma)^*$$

for all $\sigma\in Gal(K'/K)$ and $g\in S$, equality holding in $Gal(Q(X\otimes K')/K')$ hence in $Gal(Q(X\otimes K_a)/K_a)$ (via base change) where K_a is an algebraic closure of K containing K'. It is sufficient to prove that $\rho(\sigma)\in G(K_a)$ because $G(K_a)\cap Aut(X\otimes K'/K')=G(K')$. Since S is also Zariski dense in $G(K_a)$ we are reduced to proving the following:

(3.6) LEMMA. Let $C,G,X,x_\bullet$ be as in (3.4) and let $\varphi\in Aut(X/C)$. Suppose there exists a Zariski dense subset S in $G(C)$ such that

$$(\pi^{**}R_g^*)^{-1}\varphi^*(\pi^{**}R_g^*)=\varphi^*$$

for all $g\in S$, equality holding in $Gal(Q(X)/C)$. Then

$$\varphi\in Im(G(C)\longrightarrow Aut(X/C))$$

Remark. Of course we shall apply our lemma to (C,G,X,φ) replaced by $(K_a, G\otimes_C K_a, X\otimes_C K_a, \rho(\sigma))$.

Proof. By its very definition $\pi^{**}R_g^*$ is induced by a rational map $U_g: X \dashrightarrow X$ which is defined on Gx_o and for which

$$U_g(hx_o)=hgx_o \quad \text{for all} \quad h\in G(C)$$

Since $\varphi(Gx_o)\cap Gx_o\neq\phi$ there exist $g_1, g_2\in G(C)$ such that $\varphi(g_1x_o)= =g_2x_o$. Put $x_1=g_1x_o$ and $w=g_2g_1^{-1}$ hence $\varphi(x_1)=wx_1$. Now for all $g,k\in G(C)$ we have

$$U_{g_1^{-1}gg_1}(kx_1)=U_{g_1^{-1}gg_1}(kg_1x_o)=(kg_1)(g_1^{-1}gg_1)x_o=kgx_1$$

hence for all $g\in g_1Sg_1^{-1}$ we have

$$\varphi(gx_1)=\varphi(U_{g_1^{-1}gg_1}(x_1))=U_{g_1^{-1}gg_1}(\varphi(x_1))=U_{g_1^{-1}gg_1}(wx_1)=$$
$$=wgx_1=\tilde{w}(gx_1)$$

where $\tilde{w}\in \mathrm{Im}(G(C)\longrightarrow \mathrm{Aut}(X/C))$, $\tilde{w}(u):=wu$. Since φ and $\tilde{w}$ agree on a dense subset $g_1Sg_1^{-1}x_1$ of X they are equal and the lemma is proved.

(3.7) Let's prove the "easy part" 1)$\Longrightarrow$2) in (3.1) so let F/K be a strongly normal extension and find a projective Δ-model for it. Let W be a PHS for $G:=G_{F/K}$ as in (I.3.3) and choose a finite Galois extension K'/K such that $W(K')\neq\phi$. Choose a point in $W(K')$ and consider the isomorphism of G-spaces $G\otimes K'\xrightarrow{\tau} W\otimes K'$ corresponding to this point; the associated cocycle will belong to $Z^1(\mathrm{Gal}(K'/K),G(K'))$, see (3.3). By (I.3.3) we have an isomorphism $\alpha^+:Q(W)\longrightarrow F$ such that $(\alpha^+)^{-1}\delta_F\alpha^+$ commute with right translations with elements in $G(C)$. Let α^* be the composition

$$\alpha^*:Q(G\otimes K')\xrightarrow{(\tau^*)^{-1}} Q(W\otimes K')\xrightarrow{\alpha^+} K'F$$

Since $(\alpha^*)^{-1}\delta_{K'F}\alpha^*$ commute with right translations with elements in $G(C)$ we get $(\alpha^*)^{-1}\delta_{K'F}\alpha^* - \delta_{K'}{}^* \in \mathrm{Lie}_{K'}(G)$. Now by (2.4) we can find a projective variety X over C, an action $G\times X \longrightarrow X$ and a point $x_0 \in X(C)$ such that the multiplication map $\pi : G \longrightarrow X$, $\pi(g) = gx_0$ is an open immersion. Exactly as in (2.5) if we give $Q(X\otimes K')$ the structure of Δ-variety induced from $K'F$ via $Q(X\otimes K') \xrightarrow{\pi^*} Q(G\otimes K') \xrightarrow{\alpha^*} K'F$ then $X\otimes K'$ is a projective Δ-model of its function field. Now take the image of ρ into $Z^1(\mathrm{Gal}(K'/K), \mathrm{Aut}(X\otimes K'/K'))$ and still call it ρ. We have a commutative diagram

$$\begin{array}{ccccccc}
Q(X\otimes K') & \xrightarrow{\pi^*} & Q(G\otimes K') & \xrightarrow{(\tau^*)^{-1}} & Q(W\otimes K') & \xrightarrow{\alpha^+} & K'F \\
\cup & & \cup & & \cup & & \cup \\
Q(X\otimes K'/\hat{\rho}) & \xrightarrow{\sim} & Q(G\otimes K'/\hat{\rho}) & \xrightarrow{\sim} & Q(W)\otimes 1 & \xrightarrow{\sim} & F
\end{array}$$

Since $\mathcal{O}_{X\otimes K'/\hat{\rho}} = \mathcal{O}_{X\otimes K'} \cap Q(X\otimes K'/\hat{\rho})$ it follows that all derivations $\delta_{Q(X\otimes K')}$ send $\mathcal{O}_{X\otimes K'/\hat{\rho}}$ into itself hence $X\otimes K'/\hat{\rho}$ is a projective Δ-model of its Δ-function field and we are done.

(3.8) Let's prove 2) $\Longrightarrow$ 1) in (3.1). So let V be a projective Δ-model of F/K and K_a an algebraic closure of K. Then $V\otimes K_a$ is a projective Δ-model of K_aF/K_a ($K_aF := F\otimes_K K_a$), K_aF/K_a is still weakly normal by (I.3.2) and $(K_aF)^\Delta = (K_a)^\Delta = C$ where $C := K^\Delta$ by (1.4). Now apply Theorem (2.2) to K_aF/K_a and $V\otimes K_a$: we find a strongly normal extension E/K_a with $K_aF \subset E$, a descent isomorphism $\omega : V\otimes K_a \longrightarrow X\otimes_C K_a$, a connected algebraic subgroup G^* of $\mathrm{Aut}^\circ{}_{X/C} = G$, a point $p \in X(C)$ and a full G^*-primitive β of E/K_a such that

a) The orbit G^*p is Zariski open in X and the isotropy group of p in G^* is $\beta^{-1}G_{E/K_aF}\beta$ and

b) $\beta : \mathrm{Spec}\, E \longrightarrow G^*$ followed by $G^* \xrightarrow{\pi_p} G^*p \subset X$ equals $\mathrm{Spec}\, E \longrightarrow \mathrm{Spec}\, K_aF \longrightarrow V\otimes K_a \xrightarrow{\omega} X\otimes K_a \longrightarrow X$.

Since K_aF/K_a is weakly normal, $\beta^{-1}G_{E/K_aF}\beta$ is normal in $G^* = \beta^{-1}G_{E/K_a}\beta$ hence it acts trivially on G^*p by a), which forces

G_{E/K_aF} to be trivial. Hence $K_aF=E$ and $\pi := \pi_p : G^* \longrightarrow X$ is an open immersion. By b) we have a commutative diagram:

$$(*_a) \qquad \begin{array}{ccc} & Q(X\otimes K_a) & \\ \swarrow^{\pi^*} & & \searrow^{\omega^*} \\ Q(G^*\otimes K_a) & \xrightarrow{\ \beta^*\ } & K_aF \end{array}$$

Now there exists a finite Galois extension K'/K $(K'\subset K_a)$ such that ω descends to a K'-isomorphism $V\otimes_K K' \longrightarrow X\otimes_C K'$ still denoted by ω and moreover $\ell\delta\beta \in \mathrm{Lie}_{K'}(G^*)$ for all $\delta\in\Delta$. So we will have a diagram $(*')$ similar to $(*_a)$ above with K_a replaced by K'. By (3.2) there is a cocycle $\rho\in Z^1(\mathrm{Gal}(K'/K),\mathrm{Aut}(X\otimes K'/K'))$ such that ω induces an isomorphism $V \longrightarrow {}_{\rho}(X\otimes K)$ over K. We claim that ρ belongs to the image of the natural map

$$Z^1(\mathrm{Gal}(K'/K),G^*(K')) \longrightarrow Z^1(\mathrm{Gal}(K'/K),\mathrm{Aut}(X\otimes K'/K'))$$

Before proving the claim let's show how it can be used to close our proof. Consider the commutative diagram

$$\begin{array}{ccccc} Q(G^*\otimes K') & \xleftarrow{\ \pi^*\ } & Q(X\otimes K') & \xrightarrow{\ \omega^*\ } & Q(V\otimes K')=K'F \\ \cup & & \cup & & \cup \\ Q(G^*\otimes K'/\hat{\rho}) & \xleftarrow{\ \sim\ } & Q(X\otimes K'/\hat{\rho}) & \xrightarrow{\ \sim\ } & Q(V)\otimes 1 = F \end{array}$$

and give $Q(G^*\otimes K')$ the structure of Δ-field induced from $K'F$ via $\beta^* = \omega^*(\pi^*)^{-1}$. Then $Q(G^*\otimes K'/\hat{\rho})/K$ has a model which is a PHS for G^* and $G^*(C)\subset \mathrm{Gal}_\Delta(Q(G^*\otimes K')/K')\cap \mathrm{Gal}(Q(G^*\otimes K'/\hat{\rho})/K)$ since β is a G^*-primitive. By (I.3.3) $Q(G^*\otimes K'/\hat{\rho})/K$ is strongly normal hence so is F/K and Theorem (3.1) is proved modulo our claim.

To prove the claim note that by base change we get an injection

$$\mathrm{Gal}_\Delta(F/K) \longrightarrow \mathrm{Gal}_\Delta(K'F/K')=G^*(C)$$

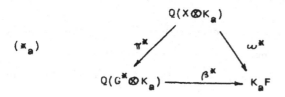

whose image S is Zariski dense in $G^*(C)$. Indeed if $H\subset G^*$ is the closure of S then by (I.3.2) we get

$$K' \subset (K'F)^H \subset (K'F)^S = F^{Gal_\Delta(F/K)} \otimes_K K' = K'$$

hence $K' = (K'F)^H$ hence by Galois correspondence $H = G^*$ (recall that $K'F/K'$ is strongly normal since it is a G^*-primitive extension !). On the other hand note that for any $g \in S$ $(\pi^*)^{-1} R_g{}^* \pi^* = (\omega^*)^{-1} g \omega^*$ and

$g(F) \subset F$ so $(\pi^{**} R_g{}^*)(Q(X \otimes K'/\hat{\rho})) \subset Q(X \otimes K'/\hat{\rho})$

and we may conclude by (3.5).

4. When are (SN) and (WN) equivalent ?

The question above was one of our main motivations for the present investigation. Of course Theorem (3.1) should be viewed as a quite precise answer to it. However we shall prove here that the (NMS) condition in that Theorem can be removed in some significant cases (it cannot be removed in general by (I.3.5)). Our main result will be:

(4.1) THEOREM. Let F/K be a Δ-function field and suppose we are in one of the following cases:

1) $tr.deg.F/K = 1$,
2) $tr.deg.F/K = 2$ and $\varkappa(F/K) \geq 0$,
3) $tr.deg.F/K = q(F/K)$ and $\varkappa(F/K) \geq 0$.

Then F/K is strongly normal if and only if it is weakly normal.

In the above statement $\varkappa(F/K)$ and $q(F/K)$ are the Kodaira dimension $\varkappa(K_aF/K_a)$ and the irregularity $q(K_aF/K_a)$ cf. (II.3.14) and (III.2.7), where K_a is an algebraic closure of K.

We should say that Kolchin proved case 1) (the "curve case") under the additional assumption that either $\varkappa(F/K) = -\infty$ (genus zero case) or K is algebraically closed [Kol_2]p.809. We shall prove it here in general. Case 2) is from the geometric point of view that of "non-ruled surfaces" and will require a careful analysis of automorphisms of surfaces fixing a given divisor globally. Case 3) will prove itself to be essentially the case of "abelian varieties" and will be easier than the preceeding one.

Before starting to analyse the three cases in Theorem (4.1) we make a general remark which is a trivial consequence of a result due to Namikawa-Ueno and Deligne [U]p.182. The remark is the following:

(4.2) LEMMA. Let K be an algebraically closed field and F a finitely generated extension of K. Suppose that $F^{Gal(F/K)}=K$. Then $\varkappa(F/K)\leq 0$.

Proof. Let V be a non-singular projective model of F/K and suppose that V has Kodaira dimension ≥ 1. Then for some $m\geq 1$ the pluricanonical map $f:V \dashrightarrow \mathbb{P}:=\mathbb{P}(H^0(V,\omega_{V/K}^{\otimes m})^{\vee})$ will have an image of dimension ≥ 1, call it W. By [U]p.182 $Gal(F/K)$ acts on $\mathbb{P}$ via some finite group and f is equivariant for this action so we get $K\subset Q(W)^{Gal(F/K)}\subset Q(V)^{Gal(F/K)}=K$ and $[Q(W):Q(W)^{Gal(F/K)}]<\infty$ hence $Q(W)=K$, contradiction. The lemma is proved.

(4.3) COROLLARY. If F/K is a weakly normal Δ-function field then $\varkappa(F/K)\leq 0$.

Proof. By (I.3.2) K_aF/K_a is weakly normal and apply (4.2).

So we see that to prove (4.1) it is sufficient to consider the cases:

2') tr.deg.$F/K=2$, $\varkappa(F/K)=0$

3') tr.deg.$F/K=q(F/K)$, $\varkappa(F/K)=0$.

Indeed case 1) for $\varkappa(F/K)=-\infty$ is due to Kolchin, loc.cit., while for $\varkappa(F/K)=0$ it reduces to case 3'). Finally note that 3') leads (at least for algebraically closed K) to abelian varieties (by Kawamata's characterisation [Kaw]).

Let's introduce some notations. Throughout (4.4)-(4.12) V will denote a non-singular projective variety over an algebraically closed field K. If D is a divisor on V the group $Aut(V,D):=\{g\in Aut(V); g^*D=D\}$ is a closed subgroup of Aut(V) and we denote by $Aut^{\circ}(V,D)$ its identity component. Now if $\lambda\in NS(V)$ put $Aut(V,\lambda)=\{g\in Aut(V); g^*\lambda=\lambda\}$; then clearly $Aut(V,\lambda)$ contains $Aut^{\circ}(V)$. Moreover the quotient $Aut(V,\lambda)/Aut^{\circ}(V)$ is finite provided λ is ample.

In dealing with case 2') from (4.3) the following will play a key role:

(4.4) PROPOSITION. Let V be a minimal non-ruled surface and D a divisor on it such that $Q(V)^{Aut(V,D)}=K$. Then either D=0 or the support of D is a disjoint union of A-D-E curves.

Recall from [BPV]p.74 that an A-D-E curve on a surface is a connected effective divisor all of whose irreducible components $D_1,\ldots,D_p$ are non-singular rational curves with selfintersection -2 and for which the intersection matrix $(D_i.D_j)$ is negative definite.

It worths noting that the case $D\neq 0$ may really occur in Proposition (4.4): take V to be the Kummer surface associated to $E\times E$ where E is an elliptic curve and take D to be the sum of the 16 distinguished (-2)-curves on V cf. [BPV]p.251. One can easily check that $Q(V)^{Aut(V,D)}=K$.

To prove Proposition (4.4) we shall prove first a series of lemmas. Throughout (4.5)-(4.11) we shall make the additional assumption that dim V=2 i.e. that V is a surface. To simplify proofs we shall sometimes tacitly assume that K has the same cardinal as $\mathbb{C}$ (and hence that it is isomorphic to $\mathbb{C}$);the general case can be easily reduced to this one.

(4.5) LEMMA. Let $\lambda \in NS(V)$ with $(\lambda.\lambda)>0$. Then $Aut(V,\lambda)/Aut^{\circ}(V)$ is a finite group.

Proof. Let $u: Aut(V,\lambda)/Aut^{\circ}(V) \longrightarrow O(NS(V),\lambda)$ be the natural representation of automorphisms into the group $O(NS(V),\lambda)$ of orthogonal automorphisms of the lattice $(NS(V),(\ .\ ))$ keeping λ fixed. Now $Ker(u) \subset Aut(V,h)/Aut^{\circ}(V)$ for some (in fact for any) ample $h \in NS(V)$, hence $Ker(h)$ is finite. Finally by the Hodge index theorem $(\ .\ )$ is negative definite on the orthogonal complement of λ in $NS(V)\otimes\mathbb{Q}$ hence $O(NS(V),\lambda)$ is also finite and we are done.

(4.6) LEMMA. Let $D \neq 0$ be a divisor on V, $Supp(D)=D_1\cup \ldots \cup D_p$ with D_i integral curves. Suppose $Aut(V,D)/Aut^{\circ}(V,D)$ is infinite. Then the intersection matrix of the divisors $D_1,\ldots,D_p$ is negative semi-definite.

Proof. Suppose there exist integers $m_1,\ldots,m_p$ such that $(\lambda.\lambda)>$ >0 where λ is the image of $m_1D_1+\ldots+m_pD_p$ in $NS(V)$ and look for a contradiction. Each $g \in Aut(V,D)$ induces a permutation of the components $D_1,\ldots,D_p$ so we have a natural group homomorphism $u: Aut(V,D) \longrightarrow S_p$ where S_p is the corresponding symmetric group. Now

$$Ker(u)= \bigcap_{i=1}^{p} Aut(V,D_i) \subset Aut(V,\lambda)$$

Since $Aut(V,D)$ meets infinitely many components of $Aut(V)$ the same will hold for $Ker(u)$ and hence for $Aut(V,\lambda)$ contradicting (4.5).

(4.7) LEMMA. Suppose V is minimal with $\varkappa(V)=0$ and $D \neq 0$ is a connected effective divisor with irreducible components $D_1,\ldots,D_p$. Suppose the intersection matrix of the curves $D_1,\ldots,D_p$ is negative semi-definite. Then either D is an A-D-E curve or there exists an elliptic fibration ([BPV] p.149) $f: V \longrightarrow B$ such that $Supp(D)$ is (set-theoretically) a fibre of f.

Proof. It is an easy consequence of [BPV] p.16 that D is either an A-D-E curve or $\mathrm{Supp}(D)$ is the support of an elliptic configuration (see [BPV] p.273 for the definition). By classification of surfaces with $\kappa=0$ [BPV] p.188 we get four cases:

Case 1: V is hyperelliptic. Let $u_i: V \longrightarrow B_i$, $i=1,2$ be two distinct elliptic fibrations and F_1, F_2 fibres of u_1, u_2 respectively ($(F_1.F_2)>0$). Since $B_2(V)=2$ (B_2=second Betti number cf. [BPV] p.148) we get that D is numerically equivalent to $a_1F_1+a_2F_2$, $a_1, a_2 \in \mathbb{Q}$. Since $(D.D) \leq 0$ we get $a_1a_2=0$ hence say $a_1=0$. Since $(D.F_2)=0$, u_2 contracts D and we are done.

Case 2: V is an Enriques surface. By [BPV] p.273 if $\mathrm{Supp}(D)=\mathrm{Supp}(E)$ with E an elliptic configuration then either $|E|$ or $|2E|$ is an elliptic pencil and we are done.

Case 3: V is a K3 surface. Again if $\mathrm{Supp}(D)=\mathrm{Supp}(E)$ as above then by Riemann-Roch $h^0(\mathcal{O}(E)) \geq 2$ and $|E|$ will give (via a possible Stein factorisation) an elliptic fibration with the desired property.

Case 4: V is an abelian surface. Then D must be an elliptic curve which may be assumed to pass through the origin of V. The quotient map $V \longrightarrow V/D$ is the desired elliptic fibration.

(4.8) LEMMA. Suppose V is minimal with $\kappa(V)=0$ and D is a connected effective divisor such that $\mathrm{Aut}(V,D)$ is infinite. Then either $D=0$ or D is an A-D-E curve or there is an elliptic fibration $f: V \longrightarrow B$ such that $\mathrm{Supp}(D)$ is (set-theoretically) a fibre of f.

Proof. If $\mathrm{Aut}^0(V,D)=1$ we are done by Lemmas (4.6) and (4.7). If $A=\mathrm{Aut}^0(V,D) \neq 1$ then by [Li$_2$] A is an abelian variety of dimension 1 and V has a structure of fibre space $f: V \longrightarrow B$ with A acting transitively on the fibres and we are done again.

(4.9) LEMMA. Let $f:P \longrightarrow S$ be a projective smooth morphism of non-singular varieties over K such that $f^{-1}(s)$ is an abelian variety for all closed points $s \in S$. Then there exist a Zariski open set $S_o \subset S$ and a finite etale morphism $S^* \longrightarrow S_o$ such that

1) $S \setminus S_o$ has codimension ≥ 2 in S,

2) $P \times_S S^* \longrightarrow S^*$ is a projective abelian scheme (i.e. a group scheme projective over S^*, see [Mum] p.115).

Remark. The above general result will be applied only in the case $\dim(S)=1, \dim(P)=2$.

Proof. The natural S-morphism $P \longrightarrow Alb^1(P/S)$ [FGA] is bijective so it is an isomorphism by ZMT. In particular $P \longrightarrow S$ is a PHS for the projective abelian scheme $A:=Alb^o(P/S) \longrightarrow S$. There exists an irreducible reduced closed subscheme T in P dominating S and such that $\dim(T)=\dim(S)$. Let $T^* \longrightarrow T$ be a desingularisation. Consider the closed set of points in S where the fibre of $p:T^* \longrightarrow S$ is not finite; clearly it has codimension ≥ 2 in S so we may remove it from S and suppose p is finite. By [Ha] p.276 p will be also flat; let n be its degree. By [Iv] p.23 there exists a section σ of the projection $(T^*/S)^{(n)} \longrightarrow S$ where $(T^*/S)^{(n)}$ denotes the relative symmetric product of T^*/S. By [Kn] p.180 the relative symmetric product $(P/S)^{(n)}$ exists as a quasi-projective scheme over S and clearly it is proper over S; the same will hold for $(A/S)^{(n)}$. Consider the morphism $\delta: P \longrightarrow S \xrightarrow{\sigma} (T^*/S)^{(n)} \longrightarrow (P/S)^{(n)}$ and denote by $\varepsilon: A \times_S P \longrightarrow P$ the action of A on P; on the fibres write $\varepsilon(a,x)=a+x$. There is a morphism $\varepsilon^n:(A/S)^n \times_S P \longrightarrow (P/S)^n$ deduced from ε which on the fibres looks as follows:

$$(a_1,\dots,a_n,x) \longmapsto (a_1+x,\dots,a_n+x)$$

Then ε^n induces a morphism $\alpha:(A/S)^{(n)} \times_S P \longrightarrow (P/S)^{(n)}$. Denote by $\beta:(A/S)^{(n)} \longrightarrow A$ the "sum" morphism and by $\gamma:A \longrightarrow A$ the multiplication by n. Then put $Z=P\times_S A\times_S (A/S)^{(n)}\times_S P$. Denote by p_1, p_{23}, a.s.o the projections of Z onto P, $A\times_S (A/S)^{(n)}$, a.s.o., and let Z_1, Z_2, Z_3 be the inverse images of the diagonals via each of the following morphisms:

$$p_1 \times (\varepsilon \circ p_{24}): Z \longrightarrow P\times P$$

$$(\beta \circ p_3)\times(\gamma \circ p_2): Z \longrightarrow A\times A$$

$$(\delta \circ p_4)\times(\alpha \circ p_{34}): Z \longrightarrow (P/S)^{(n)}\times (P/S)^{(n)}$$

By separatedness, Z_1, Z_2, Z_3 are closed subschemes of Z; put $\hat{S}= p_1(Z_1\cap Z_2\cap Z_3)$. Since p_1 is proper $\hat{S}$ is closed and we consider it as a subscheme of P with its reduced structure. In down to earth terms $\hat{S}$ has the following description: for each fibre F of $P\longrightarrow S$ we fix an arbitrary closed point $x\in F$ and consider the Albanese map $\varphi_x: F\longrightarrow \mathrm{Alb}(F)$ for which $\varphi_x(x)=0$; of course φ_x is an isomorphism. Using σ one can think about $F\cap T$ as a 0-cycle $\sum_{i=1}^{n} x_i$ on F. Let $a_1,\dots,a_{n^{2g}} \in \mathrm{Alb}(F)$ be the solutions of the equation

$$na=\sum_{i=1}^{n}\varphi_x(x_i)$$

where $g=\dim(F)$ and for any $j=1,\dots,n^{2g}$ put $y_j=\varphi_x^{-1}(a_j)$. The correspondence

$$\sum_{i=1}^{n} x_i \longmapsto \sum_{j=1}^{n^{2g}} y_j$$

will not depend on the choice of $x\in F$. Furthermore $\hat{S}$ is spread out by the cycles $\sum y_j$ as F runs through all fibres of $P\longrightarrow S$. So the cardinals of the fibres of $\hat{S}\longrightarrow S$ at the closed points of S are the same (equal to n^{2g}). Let $S^*\longrightarrow \hat{S}$ be a desingularisation. Removing from S a codimension 2 closed subset we may suppose that

$S^* \longrightarrow S$ is finite and hence by [Ha] p.276 also flat. Moreover the cardinals of the fibres of $S^* \longrightarrow S$ at the closed points of S are still the same. We conclude that $S^* \longrightarrow S$ is étale. Now $P \times_S S^*/S^*$ is a PHS for the projective abelian scheme $A \times_S S^*/S^*$ and has a section hence it is isomorphic to $A \times_S S^*/S^*$ [Mi] p.120 and we are done.

(4.10) LEMMA. Let $f: V \longrightarrow \mathbb{P}^1$ be an elliptic fibration. Suppose f has at most two degenerate fibres and there exists an infinite set $S \subset \mathbb{P}^1$ and an elliptic curve E such that $f^{-1}(x) \simeq E$ for all $x \in S$. Then V is ruled.

Proof. Let $\mathcal{L} \in \mathrm{Pic}(V)$ be ample relative to f and put $F_b = f^{-1}(b)$ scheme-theoretically for all $b \in B = \mathbb{P}^1$. Let $B_o = \{b \in B;\ F_b$ is smooth$\}$ and $V_o = f^{-1}(B_o)$. By Lemma (4.9) there is an étale finite morphism $B^* \longrightarrow B_o$ such that $V^* := V_o \times_{B_o} B^* \longrightarrow B^*$ is a projective abelian scheme. Now by [Popp] lecture 10, there is an étale finite morphism $B^{**} \longrightarrow B^*$ such that $V^{**} := V^* \times_{B^*} B^{**} \longrightarrow B^{**}$ has a level n structure $(n \geq 3)$ and hence the latter morphism is obtained via base change from the universal family $U \longrightarrow M$ of projective abelian schemes of dimension 1 with polarisation of some fixed degree and with level n structure, via a unique classifying morphism $e: B^{**} \longrightarrow M$. Since each elliptic curve carries finitely many level n structures, $e(S)$ is a finite set so e must be constant hence $V^{**} \simeq E \times B^{**}$. We shall be done if we prove that B^{**} is a rational curve. But B_o is either $\mathbb{C}$ or $\mathbb{C} \setminus \{0\}$ hence $\pi_1(B_o)$ is either 0 or $\mathbb{Z}$ hence any étale finite covering of B_o is either the identity of B_o or the map $\mathbb{C} \setminus \{0\} \longrightarrow \mathbb{C} \setminus \{0\}$, $z \longmapsto z^k$, $k \geq 1$. Consequently B^{**} is an open subset of $\mathbb{P}^1$ and we are done.

(4.11) Proof of Proposition (4.4). By Lemma (4.2) we have $\varkappa(V) = 0$.

Suppose $D \neq 0$ and let $D_1, \ldots, D_p$ be the irreducible components of D. As in Lemma (4.6) we have a group homomorphism $u:\mathrm{Aut}(X,D) \longrightarrow S_p$ Clearly $[Q(V)^{\mathrm{Ker}(u)}:Q(V)^{\mathrm{Aut}(V,D)}] < \infty$ hence $Q(V)^{\mathrm{Ker}(u)}=K$, in particular if C is a connected component of $\mathrm{Supp}(D)$ with its reduced structure then $Q(V)^{\mathrm{Aut}(V,C)}=K$. Suppose C is not an A-D-E curve and look for a contradiction. By Lemma (4.8) there exists an elliptic fibration $f:V \longrightarrow B$ such that $C=(F_b)_{red}$ for some $b \in B$ (where as usual $F_x=f^{-1}(x)$, $x \in B$). Let $b_1, \ldots, b_q \in B$ be the set of all points in B at which f has a degenerate fibre. We claim that for any $g \in \mathrm{Ker}(u)$ we have a commutative diagram

$$\begin{array}{ccc} V & \xrightarrow{\quad g \quad} & V \\ {\scriptstyle f}\downarrow & & \downarrow{\scriptstyle f} \\ B & \xrightarrow{\quad \tilde{g} \quad} & B \end{array}$$

with $\tilde{g} \in \mathrm{Aut}(B)$. Since $f_* \mathcal{O}_V = \mathcal{O}_B$ it is sufficient to check the existence of the above diagram set-theoretically. Take $x \in B$ and let's show that $g(F_x)$ is contracted by f. We have $F_x \approx F_b$ ($\approx$ denoting the algebraic equivalence) hence $g(F_x) \approx g(F_b)=F_b$ hence $(g(F_x).F_b)=0$ and we are done. Now we got a group homomorphism $\mathrm{Ker}(u) \longrightarrow \mathrm{Aut}(B)$ whose image H is contained in $\mathrm{Aut}(B,b) \cap \mathrm{Aut}(B,b_1+\ldots+b_q)$. As in Lemma (4.6) we get a group homomorphism $v:H \longrightarrow S_q$ so $\mathrm{Ker}(v) \subset \mathrm{Aut}(B,b) \cap \mathrm{Aut}(B,b_1) \cap \ldots \cap \mathrm{Aut}(B,b_q)$. Now

$$K \subset Q(B)^H \subset Q(V)^{\mathrm{Ker}(u)}=K \qquad \text{and} \qquad [Q(B)^{\mathrm{Ker}(v)}:Q(B)^H] < \infty$$

so $\mathrm{Ker}(v)$ must be infinite. Consequently $B=\mathbb{P}^1$ and if $B_s=\{b\} \cup \{b_1, \ldots, b_q\}$ then $1 \leq \# B_s \leq 2$, in particular f has at most 2 degenerate fibres. To find a contradiction it is sufficient by Lemma (4.10) to find an infinite set $S \subset B$ and an elliptic curve E such that $F_x \simeq E$ for all $x \in S$.

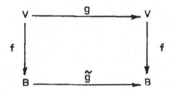

If $\#B_s=1$ take a sequence $(g_n)_n$ of distinct elements in Ker(v) and choose an affine coordinate z such that B_s is given by $z=\infty$ and $g_n(z)=a_nz+b_n$, $a_n\in K^*$, $b_n\in K$. Supposing K uncountable there exists $z_0\in K$ not belonging to the field generated by all a_n's and b_n's. Then $g_n(z_0)$ is a sequence of distinct elements in B and $F_{g_n(z_0)}\simeq F_{z_0}$ for all n (we suppose g_1=identity).

If $\#B_s=2$ choose g_n and z such that B_s consists of the points $z=0$ and $z=\infty$ and such that $g_n(z)=a_nz$, $a_n\in K^*$. We conclude in the same way as above.

Now in dealing with case 3') in Theorem (4.1) cf. (4.3) we will need the following analog of Proposition (4.4) for abelian varieties:

(4.12) PROPOSITION. Let V be an abelian variety and D a divisor on it such that $Q(V)^{Aut(V,D)}=K$. Then $D=0$.

Proof. Suppose $D\neq 0$. Considering permutations induced by Aut(V, D) on the set of irreducible components of Supp(D) we may suppose that D is an integral divisor. Identify V with $Aut^0(V)$. Then $W:=V\cap Aut(V,D)$ is a closed subgroup of V distinct from V. Put $A=V/W$ and let $p:V\longrightarrow A$ be the natural projection and $E=p(D)$. Since W is a normal subgroup in Aut(V,D), Aut(V,D) will still act on A via some subgroup of Aut(A,E). Note that there is no $a\in A\setminus\{0\}$ such that $E+a=E$ hence by [La] pp.87 and 94 E is ample. Since

$$K\subset Q(A)^{Aut(A,E)}\subset Q(V)^{Aut(V,D)}=K$$

we shall be done if we check that Aut(A,E) is finite. Now ampleness of E implies that Aut(A,E) has finitely many components. On the other hand its connected component is contained in $A\cap Aut(A,E)$ which is finite by [La] p.96 and we are done.

(4.13) Proof of Theorem (4.1). By the remark made after Corollary

(4.3) we are reduced to considering cases 2') and 3').

Case 3'). So we suppose tr.deg.$F/K=q(F/K)$, $\varkappa(F/K)=0$ and F/K weakly normal and we want to prove that F/K is strongly normal. Start with any non-singular projective model W of F/K and consider the natural morphism $W \longrightarrow W_1:=\mathrm{Alb}^1(W/K)$. By [FGA] W_1 is a PHS for $\mathrm{Alb}^0(W/K)$ hence it is projective and non-singular. By compatibility with base change and by [Kaw] the morphism $W\otimes K_a \longrightarrow W_1\otimes K_a = \mathrm{Alb}^1(W\otimes K_a/K_a)$ is birational hence so is the original morphism. Hence W_1 is still a projective model of F/K. We claim W_1 is a Δ-model and this will close the proof by (3.1). By (I.2.5) it is sufficient to check that $W_1\otimes K_a$ is a Δ-model. Let D be the divisor of movable singularities on $W_1\otimes K_a$. Notations being as in (II.4.6) we have $\mathrm{Gal}_\Delta(K_aF/K_a)=\mathrm{Aut}_\Delta(W_1\otimes K_a/K_a)\subset \mathrm{Aut}(W_1\otimes K_a,D)$. By (I.3.2) K_aF/K_a is weakly normal hence we get $(K_aF)^{\mathrm{Aut}(W_1\otimes K_a,D)}=K_a$. By (4.12) $D=0$ and we are done.

Case 2') The hypothesis is tr.deg.$F/K=2$, $\varkappa(F/K)=0$ and F/K weakly normal and again we want to prove that F/K is strongly normal. By Case 3') we shall be done if we prove that $q(F/K)=2$. Take a non-singular projective minimal model V of K_aF/K_a and let D be the divisor of movable singularities on V. Again we have $\mathrm{Gal}_\Delta(K_aF/K_a)=\mathrm{Aut}_\Delta(V/K_a)\subset \mathrm{Aut}(V,D)$ so $K_aF^{\mathrm{Aut}(V,D)}=K_a$. By (4.4) either $D=0$ or $\mathrm{Supp}(D)$ is a disjoint union of A-D-E curves. The second alternative is impossible by (I.2.9). By (3.1) K_aF/K_a is strongly normal. By (I.3.19) it is an abelian extension in particular $q(F/K)=2$ and we are done.

CHAPTER IV. COMPLEMENTS.

In this chapter we discuss relationship between our theory and "classical beings". There will be few results and many comments through which we hope to convince the reader that Kolchin's theory as well as the theories developed in [Mtd] and in the present book are not abstract and formal but on the contrary quite effective.

1. Special values of automorphic functions.

Classical theorems in algebraic number theory (going back to Kronecker and Weber) assert that certain Galois extensions F/K of algebraic number fields with commutative Galois groups can be generated by "special values" of automorphic functions (such as the exponential function exp, the elliptic modular function J, the Weierstrass $\wp$ function a.s.o.); more precisely that F has the form $F=K(\varphi(z))$ where $z\in K$ and $\varphi=(\varphi_1,\ldots,\varphi_r)$, φ_j being certain automorphic functions (cf. for instance [Sh_1] pp. 1 and 15). Striking analogs of this type of results can be proved in the frame of Kolchin's differential Galois theory. A result of this kind was stated by Kolchin in [Kol_3] for abelian extensions of fields of meromorphic functions. We shall give here a proof of a slightly more general result in which we consider strongly normal extensions with arbitrary commutative Galois group. The automorphic functions arrising in this way will be the so called "coordinates of the exponential map" of a commutative algebraic group [Wald]; they include the usual exponential and the abelian functions (attached to a fixed abelian lattice) but do not include functions like the elliptic modular function and its higher dimensional generalisations [Sh_1]. To bring such functions into the picture one has to consider also Δ-fields of meromorphic functions"with parameters", see (1.2) and (1.7) below.

The moral of this § should be that our main results in the previous chapter (especially Theorems (III.3.1) and (III.4.1)) become in the analytic case criteria for a system of algebraic differential equations to be"linearisable by means of abelian functions" (in a sense which will be discussed below).

(1.1) Start with a connected commutative algebraic group G of dimension n over the complex field $\mathbb{C}$. Then G^{an} is a complex Lie group and it is well known that the exponential map $\exp:\mathrm{Lie}(G)\longrightarrow G^{an}$ identifies G^{an} with the quotient $\mathrm{Lie}(G)/\Omega$ where Ω is a discrete subgroup of Lie(G). Let's identify $\mathrm{Lie}(G)$ with $\mathbb{C}^n$.Then we have an embedding

$$(*)\qquad Q(G)\longrightarrow M(G^{an})\xrightarrow{\ \exp^*\ } M(\mathbb{C}^n)$$

where $M(X)$=field of meromorphic functions on the analytic manifold X. The image of $Q(G)$ in $M(\mathbb{C}^n)$ will be denoted by $A(G)$. By the way the structure of the field $A(G)$ is not very well understood although some beatiful properties of it can be proved $[S_1]$. Now we have an isomorphism $\mathrm{Lie}(\mathbb{C}^n)\simeq \mathrm{Lie}(\mathrm{Lie}(G))\xrightarrow{\ d(\exp)\ }\mathrm{Lie}(G)$. If $z_1,\dots,z_n$ are natural coordinates on $\mathbb{C}^n$ then the basis $\partial/\partial z_1,\dots,\partial/\partial z_n$ of $\mathrm{Lie}(\mathbb{C}^n)$ will correpond to a basis $\theta_1,\dots,\theta_n$ of $\mathrm{Lie}(G)$. Put $\Lambda=\{\lambda_1,\dots,\lambda_n\}$ and view $Q(G)$ and $M(\mathbb{C}^n)$ as Λ-fields by putting $\lambda_{jQ(G)}=\theta_j$ and $\lambda_{jM(\mathbb{C}^n)}=\partial/\partial z_j$; then the embedding $(*)$ is a Λ-field extension hence $A(G)$ is a Λ-subfield of $M(\mathbb{C}^n)$.

(1.2) Let R be a region in $\mathbb{C}^m\times\mathbb{C}^p$, $m\geqslant 1$, $p\geqslant 0$, $w_1,\dots,w_m$ coordinates in $\mathbb{C}^m$, $u_1,\dots,u_p$ coordinates in $\mathbb{C}^p$ and view $M(R)$ as a Δ-field, $\Delta=\{\delta_1,\dots,\delta_m\}$, $\delta_{jM(R)}=\partial/\partial w_j$, $1\leq j\leq m$.

We shall call $u_1,\dots,u_p$ parameters. Throughout (1.2)-(1.6) we shall consider (as we did in the Introduction) only the case $p=0$, i.e.

the case of "no parameters".

Now let R' be a subregion of R, let K be a Δ-subfield of M(R) containing $\mathbb{C}$ and $\beta_1,\ldots,\beta_n$ holomorphic functions on R' such that if $\beta=(\beta_1,\ldots,\beta_n):R'\longrightarrow \mathbb{C}^n$ then $\varphi(\beta):=\varphi\circ\beta$ is defined (as a meromorphic function on R') for all $\varphi\in A(G)$ and such that $\delta_j\beta_i=\partial\beta_i/\partial w_j\in K$ for all $1\le i\le n$, $1\le j\le m$. To the data K, β, G we may associate a Δ-field extension F/K (with F a Δ-subfield of M(R')) as follows: we let F be the subfield of M(R') generated by K and all functions of the form $\varphi(\beta)$ where $\varphi\in A(G)$. By abuse we write $F=K(\varphi(\beta))$. Clearly F/K is finitely generated as a field extension.

(1.3) PROPOSITION. Let G be a connected commutative algebraic group over $\mathbb{C}$ of dimension n and let F/K be a Δ-field extension of Δ-subfields of M(R), $R\subset\mathbb{C}^m$ with $\mathbb{C}\subset K$. Suppose F/K is a full G-primitive extension. Then $F=K(\varphi(\beta))$ for some β as in (1.2).

(1.4) Before proving the above proposition let's show how it applies to strongly normal extensions (and hence by our theory in (III.3) and (III.4) to weakly normal extensions !). Suppose F_o/K_o is a Δ-function field where F_o is a Δ-subfield of some $M(R_o)$, $R_o\subset\mathbb{C}^m$ and suppose F_o/K_o is strongly normal with commutative Galois group $G:=G_{F_o/K_o}$. By (I.3.20) there exists a finite extension K_1/K_o such that K_1F_o/K_1 is a full G-prmitive extension. Now by the "analytic theorem of zeroes" due to Ritt [Ri] p.176 there is a K_o-embedding of K_1 into some $M(R_1)$, $R_1\subset R_o$. Consequently K_1 and $K_1F_o=F_o\otimes_{K_o}K_1$ are subfields of $M(R_1)$ and we may apply Proposition (1.3) to the extension K_1F_o/K_1.

(1.5) Proof of (1.3). Let $\alpha:\mathrm{Spec}\,F\longrightarrow G\otimes K$ be a full G-primitive of F/K. By (I.3.17) α induces an isomorphism $\sigma:F\longrightarrow Q(G\otimes K)$ such that

$$\sigma\delta_{jF}\sigma^{-1}=\delta_{jK}^{*}+\sum_{k=1}^{n}\theta_k\otimes a_{jk}\qquad \text{where}\quad a_{jk}\in K$$

Relations $[\theta_i,\theta_j]=0$ and $[\sigma\delta_p\sigma^{-1},\sigma\delta_q\sigma^{-1}]=0$ immediately imply

$$\partial a_{jk}/\partial w_p=\partial a_{pk}/\partial w_j \qquad \text{for all } j,p,k$$

Now let Spec $A\subset G$ be a Zariski open set; write $A=\mathbb{C}[T_1,\dots,T_N]/J=\mathbb{C}[u_1,\dots,u_N]$, J being a prime ideal in $\mathbb{C}[T_1,\dots,T_N]$. Let ψ_q be the images of $u_q\otimes 1\in Q(A\otimes K)=Q(G\otimes K)$ in $M(R)$ via $\sigma^{-1}:Q(G\otimes K)\longrightarrow F\subset M(R)$. Since $\theta_k(A)\subset A$ there exist polynomials $P_{qk}\in\mathbb{C}[T_1,\dots,T_N]$ such that

$$\theta_k u_q=P_{qk}(u_1,\dots,u_N) \quad \text{for all } k,q$$

We have

$$\partial\psi_q/\partial w_j=\sigma^{-1}\Big(\sum_{k=1}^{n}\theta_k u_q\otimes a_{jk}\Big)=\sum_{k=1}^{n}a_{jk}P_{qk}(\psi_1,\dots,\psi_N)$$

for all j,q. Furthermore $f(\psi_1,\dots,\psi_N)=0$ for all $f\in J$. By the holomorphic Poincaré lemma [GH] p.448 there exists a subregion $R'\subset R$ and functions $b_1,\dots,b_n$ holomorphic in R' such that $\partial b_k/\partial w_j=a_{jk}$ for all j and k. Choose $w^0=(w_1^0,\dots,w_m^0)\in R'$ such that all b_k's and ψ_q's are holomorphic at w^0. Since $y^0:=(\psi_1(w^0),\dots,\psi_N(w^0))$ is a $\mathbb{C}$-point of Spec $A\subset\mathbb{C}^N$ we may write $\psi_q(w^0)=u_q(y^0)$ for all q. Take $x^0\in\mathbb{C}^n$ such that $\exp(x^0)=y^0$ and define

$$\eta_q(w)=\varphi_q(b_1(w)-b_1(w^0)+x_1^0,\dots,b_n(w)-b_n(w^0)+x_n^0)$$

where $\varphi_q=u_q\circ\exp$; clearly η_q are holomorphic at w^0 and $\eta_q(w^0)=\psi_q(w^0)$. Shrinking R' we may suppose that η_q are holomorphic on R'. Now using the identity

$$(\partial/\partial z_k)(u_j\circ\exp)=(\theta_k u_j)\circ\exp$$

we see that

$$\partial\eta_q/\partial w_j=\sum_{k=1}^{n}a_{jk}P_{qk}(\eta_1,\dots,\eta_N) \quad \text{for all } j,q$$

One obtains inductively that the Taylor expansions of η_q and ψ_q around w^o coincide hence $\eta_q=\psi_q$. Now $\varphi_1,\dots,\varphi_N$ generate $A(G)$ over $\mathbb{C}$, $\psi_1,\dots,\psi_N$ generate F over K and $\psi_q=\varphi_q(\beta)$ where $\beta=(\beta_1,\dots,\beta_n)$, $\beta_i=b_i-b_i(w^o)+x_i^o$ so we are done.

(1.6) The following consequence of (1.3),(1.4),(I.3.4) and Chevalley's structure theorem [Ro] worths being noted. Suppose F/K is a strongly normal Δ-function field, $\mathbb{C}\subset K\subset F\subset M(R)$ for some region $R\subset\mathbb{C}$ Then there exists a linear system S of ordinary differential equations whose coefficients have the form

$$\psi(\dots,\gamma_i,\dots,\varphi_j(\int\alpha),\dots)$$

(where ψ is an algebraic function, $\gamma_i\in K$, φ_j are abelian functions and α is a function algebraic over K) such that any function in F is a rational combination (with coefficients algebraic over K) of solutions of S. One can make of course more precise statements (holding also in the partial case) but the above formulation already suggests what one means by saying that strong normality implies "linearisation by means of abelian functions".

(1.7) As we announced in the beginning of this §, by allowing p to be >0 in (1.2) (in other words by allowing "parameters") one can bring modular functions (such as the J function) into the setting of strongly normal extensions. To give a hint of what we mean by this we shall analyse an example which in some sense is the simplest non-trivial one, see (1.8) below. Start by recalling some well known formulas from classical theory of elliptic functions [SZ]. Consider the holomorphic functions

$$g_2(\omega,\omega')=60\sum{}'(m\omega+n\omega')^{-4}$$

$$g_3(\omega,\omega')=140\sum{}'(m\omega+n\omega')^{-6}$$

$$\Delta(\omega,\omega')=g_2{}^3(\omega,\omega')-27g_3{}^2(\omega,\omega')$$

where $\omega/\omega' \notin \mathbb{R}$,

$$J(\tau)=g_2{}^3(1,\tau)/\Delta(1,\tau)=1+27g_3{}^2(1,\tau)/\Delta(1,\tau)$$

where $\operatorname{Im}\tau>0$ and the meromorphic function

$$\wp(z,\omega,\omega')=z^{-2}+\sum{}'((z-m\omega-n\omega')^{-2}-(m\omega+n\omega')^{-2})$$

where $\sum{}'$ denotes summation over all pairs of integers $(m,n)\neq(0,0)$. The following identity holds:

$$(\partial\wp/\partial z)^2=4\wp^3-g_2\wp-g_3 \qquad \text{for all } z,\omega,\omega'$$

Note also that J^{-1} is an infinitely valued analytic function with critical points $0,1,\infty$ only. We shall denote by J_α^{-1} an arbitrary element belonging to J^{-1} defined on a disc $D(\alpha,r_\alpha)$ with center α and radius r_α satisfying $D(\alpha,r_\alpha)\cap\{0,1\}=\emptyset$. Put

$$T_\alpha(a,b)=J_\alpha{}^{-1}(a^3/a^3-27b^2)$$

which is defined for $|\alpha-(a^3/a^3-27b^2)|<r_\alpha$ (this implies in particular that $a\neq 0$, $b\neq 0$). Put

$$W_\alpha(a,b)=ab^{-1}g_2(1,T_\alpha(a,b))(g_3(1,T_\alpha(a,b)))^{-1}$$

$$W'_\alpha(a,b)=T_\alpha(a,b)W_\alpha(a,b)$$

By [SZ] p.403 we have

$$g_2(W_\alpha(a,b),W'_\alpha(a,b))=a$$

$$g_3(W_\alpha(a,b),W'_\alpha(a,b))=b$$

whenever $|\alpha-(a^3/a^3-27b^2)|<r_\alpha$. Now define for $\alpha\in\mathbb{C}\setminus\{0,1\}$

$$\varphi_\alpha(z,a,b)=\wp(z,W_\alpha(a,b),W'_\alpha(a,b))$$

(1.8) PROPOSITION. Let R_o be a region in $\mathbb{C}\times\mathbb{C}^p$, let w be a coordinate in $\mathbb{C}$ and $u=(u_1,\ldots,u_p)$ coordinates in $\mathbb{C}^p$. View $M(R_o)$ as an ordinary Δ-field with derivation $\delta=\partial/\partial w$ and let F_o/K_o be a Δ-function field with no movable singularity of transcendence degree 1 and genus 1 with j-invariant different from 0 and 1, such that $\mathbb{C}\subset K_o$ and F_o is a Δ-subfield of $M(R_o)$. Then there is a subregion $R_1\subset R_o$, a finite extension K_1/K_o, $K_1\subset M(R_1)$, functions $\beta_1,\beta_2,\beta_3\in M(R_1)$ and a complex number $\alpha\notin\{0,1\}$ such that: $\partial\beta_1/\partial w\in K_1$, $\partial\beta_2/\partial w=\partial\beta_3/\partial w=0$ and either K_1F_o/K_1 is split or

$$K_1F_o=K_1(\varphi_\alpha(\beta_1,\beta_2,\beta_3),(\partial\varphi_\alpha/\partial z)(\beta_1,\beta_2,\beta_3))$$

Proof. Arguments are similar to those in (1.3). We give them for the sake of completeness. By [Mtd] p.37 or by (II.1.25) there is a finite extension K_1 of K_o such that either K_1F_o/K_1 is split or $K_1F_o=K_1(\psi,\delta\psi)$ where

$$(*)\qquad (\delta\psi)^2=b^2(4\psi^3-\beta_2\psi-\beta_3)$$

with $b\in K_1\setminus\{0\}$, $\beta_2,\beta_3\in(K_1)^\Delta$, $\beta_2^3-27\beta_3^2\neq 0$. By [Ri] p.176 one can assume that K_1F_o is a Δ-subfield of some $M(R_1)$, $R_1\subset R_o$. Suppose K_1F_o/K_1 is non-split hence $\partial\psi/\partial w\neq 0$. Hence there exists a w^o such that $u\longmapsto(\partial\psi/\partial w)(w^o,u)$ does not vanish identically on $D=\{u\in\mathbb{C};(w^o,u)\in R_1\}$. Choose $u^o\in\mathbb{C}^p$ such that $(w^o,u^o)\in R_1$ and $\beta_2(u^o)\beta_3(u^o)(\partial\psi/\partial w)(w^o,u^o)\neq 0$ and put

$$\alpha=\beta_2^3(u^o)/\beta_2^3(u^o)-27\beta_3^2(u^o)$$

Then clearly $W_\alpha(\beta_2(u),\beta_3(u))$ is defined for u near u^o and the same will hold for W'_α. Moreover $b(w^o,u^o)\neq 0$. Now since

$$((\partial\psi/\partial w)(w^o,u)/b(w^o,u))^2=\psi^3(w^o,u)-\beta_2(u)\psi(w^o,u)-\beta_3(u)$$

it follows that for any $u\in D$ there is a complex number $t(u)\in\mathbb{C}$

such that

$$\wp(t(u),W_\alpha(\beta_2(u),\beta_3(u)),W'_\alpha(\beta_2(u),\beta_3(u)))=\psi(w^0,u)$$

$$\partial\wp/\partial z(t(u),W_\alpha(\beta_2(u),\beta_3(u)),W'_\alpha(\beta_2(u),\beta_3(u)))=(\partial\psi/\partial w(w^0,u))/b(w^0,u)$$

because $\wp$ and $\wp'$ give a parametrisation of the elliptic curves corresponding to each u. Now we claim that one can replace the function $u\longmapsto t(u)$ by a function $u\longmapsto \gamma(u)$ holomorphic around u^0. Indeed using the fact that $\partial\psi/\partial w(w^0,u^0)\neq 0$ we see that one can apply the implicit function theorem for the first equation above so there is a holomorphic function $u\longmapsto\gamma(u)$ around u^0 such that $\gamma(u^0)=t(u^0)$ and the first equation holds with t replaced by γ. Let's show that γ is good also for the second equation. It is certainly good for the second equation with both terms squared. So if it is not good for the second equation then it is good for the second equation with the right hand side multiplied by -1. Taking $u=u^0$ we get $\partial\psi/\partial w(w^0,u^0)=0$ contradiction. Now solve the equation

$$\begin{cases}\partial\beta_1/\partial w(w,u)=b(w,u)\\ \beta_1(w^0,u)=\gamma(u)\end{cases}$$

around (w^0,u^0) and replace R_1 by some neighbourhood of (w^0,u^0) such that $\beta_1\in M(R_1)$. Consider the system

$$\begin{cases}\partial X/\partial w=Y\\ \partial Y/\partial w=b(w,u)(3Y^2-\beta_2(u))+\partial b/\partial w(w,u)Y^{-1}(Y^3-\beta_2Y-\beta_3)\end{cases}$$

Note that both

$$\begin{cases}X_1=\psi\\ Y_1=\partial\psi/\partial w\end{cases}\quad\text{and}\quad\begin{cases}X_2=\wp_\alpha(\beta_1,\beta_2,\beta_3)\\ Y_2=b\,\partial\wp_\alpha/\partial z(\beta_1,\beta_2,\beta_3)\end{cases}$$

are solutions for this system around (w^0,u^0). On the other hand by construction $X_1(w^0,u)=X_2(w^0,u)$ and $Y_1(w^0,u)=Y_2(w^0,u)$. We conclude that $X_1=X_2$ and $Y_1=Y_2$ and we are done.

(1.9) Remark. The case of j-invariant 0 or 1 in (1.8) may be treated in the same way (in fact it is simpler since "no moduli" of elliptic curves will appear).

2. Analytic foliations versus differential algebra.

(2.1) Classically the objects one starts with are systems of partial algebraic differential equations with meromorphic coefficients (call them simply Δ-systems). By these we mean the following. Take a region $R \subset \mathbb{C}^m \times \mathbb{C}^p$ take $\Delta=\{\delta_1,\dots,\delta_m\}$ as in (1.2), take Δ-indeterminates $y_1,\dots,y_r$, consider Δ-polynomials with coefficients in $M(R)$: $f_1,\dots,f_k \in M(R)\{y_1,\dots,y_r\}$ [Kol_1]p.69 and consider the "system"

$$(2.1.1) \qquad f_j(y_1,\dots,y_r)=0 \qquad 1 \leq j \leq k$$

By a solution of such a Δ-system one understands an r-uple $\beta=(\beta_1,\dots,\beta_r)$ of meromorphic functions on a subregoin R' of R such that $f_j(\beta_1,\dots,\beta_r)=0$ for $1 \leq j \leq k$. Coordinates of $\mathbb{C}^p$ should be viewed as "parameters" of the Δ-system. For simplicity we shall assume in this § that $p=0$.

Now there are two points of view on Δ-systems we would like to compare:

1) the foliation-theoretic standpoint (cf. [GS],[J_1] going back to Fuchs, Poincaré, Painlevé,...)

2) the differential algebraic stanpoint (cf. [Ri], [Kol_1], [Mtd] and the present book).

One moral will be that there exist problems (and these are precisely the kind of problems we were essentially dealing with) in which differential algebra leads to significant results while the foliation-theoretic standpoint turns out to be helpless.

In (2.2) and (2.3) below we shall consider for simplicity only Δ-systems of the form:

$$(2.1.2)\qquad \begin{aligned} &F(y,\delta y,\ldots,\delta^d y)=0\\ &(\partial F/\partial y_d)(y,\delta y,\ldots,\delta^d y)\neq 0\end{aligned}$$

where $F\in M(R)[y_0,\ldots,y_d]$ is an absolutely irreducible polynomial in $d+1$ variables with coefficients in $M(R)$, $R\subset\mathbb{C}$ and $\delta=d/dw$ with w a coordinate in R. The second inequality should be interpreted of course as an equation of the form $x(\partial F/\partial y_d)-1=0$ where x is a new Δ-indeterminate. We shall always suppose (and this is possible by shrinking R) that all coefficients of F are holomorphic on R.

(2.2) Start with the foliation-theoretic standpoint. Recall first that given a complex analytic manifold X, a foliation on X means an integrable subbundle $\mathcal{F}$ of the tangent bundle T_X of X (where integrability means $[\mathcal{F},\mathcal{F}]\subset\mathcal{F}$; by Frobenius' theorem $[J_1]$ p.200 integrability is equivalent to $\mathcal{F}$ being locally isomorphic to T_Y for some submanifold $Y\subset X$. Such an Y is called a local integral subvariety of $\mathcal{F}$). By $\dim(\mathcal{F})$ one understands the rank of the bundle $\mathcal{F}$. Now there is a natural way of associating to any Δ-system an analytic manifold X, an analytic map $\varphi:X\longrightarrow R\subset\mathbb{C}^m$ and a foliation $\mathcal{F}$ on X of dimension m transverse to φ (in the sense that the tangent map $T_x\varphi:T_xX\longrightarrow T_{\varphi(x)}R$ maps $\mathcal{F}_x$ isomorphically onto $T_{\varphi(x)}R$). For simplicity we shall describe this construction only for the Δ-systems of the form (2.1.2).

Put

$$F(y_0,\ldots,y_d)=\sum a_{\alpha_0\ldots\alpha_d}(w)y_0^{\alpha_0}\ldots y_d^{\alpha_d}=u(w,y_0,\ldots,y_d)$$

with u holomorphic in $R\times\mathbb{C}^{d+1}$ and consider the analytic manifold X defined in $R\times\mathbb{C}^{d+1}$ by the equations:

$$u(w,y_0,\ldots,y_d)=0$$

$$\partial u/\partial y_d(w,y_0,\ldots,y_d)\neq 0$$

It comes equiped with a natural projection $\varphi:X\longrightarrow R$ whose fibres are algebraic locally closed subvaritties in $\mathbb{C}^{d+1}$. Now one can consider the 1-dimensional foliation in $U=(R\times\mathbb{C}^{d+1})\setminus\{\partial u/\partial y_d=0\}$ given by the system of Pfaff equations

$$dy_0-y_1dw=0$$

$$\cdots\cdots\cdots\cdots$$

$$dy_{d-1}-y_ddw=0$$

$$dy_d-hdw=0$$

where $h=-(\partial u/\partial y_d)^{-1}(\partial u/\partial w+y_1(\partial u/\partial y_0)+\ldots+y_d(\partial u/\partial y_{d-1}))$. At each point of U the foliation is generated by the vector field

$$\theta=\partial/\partial w+y_1(\partial/\partial y_0)+\ldots+y_d(\partial/\partial y_{d-1})+h(\partial/\partial y_d)$$

and since $\langle du,\theta\rangle=0$ we get an induced foliation $\mathcal{F}$ on X; moreover $(T\varphi)(\theta)=\partial/\partial w$ hence $\mathcal{F}$ is transverse to φ. Finally given a solution β of (2.1.2) holomorphic on some $R'\subset R$ one immediately constructs a local integral subvariety of $\mathcal{F}$ by taking the image of the holomorphic map $R'\longrightarrow X$, $w\longmapsto(w,\beta(w),(d\beta/dw)(w),\ldots,(d^d\beta/dw^d)(w))$.

(2.3) On the other hand from the algebraic standpoint if we are given a Δ-system as in (2.1.2) then one defines K to be the Δ-subfield of $M(R)$ generated over $\mathbb{C}$ by all coefficients $a_{\alpha_0\ldots\alpha_d}$ of F (i.e. the field generated by these coefficients together with all their higher derivatives) and one puts $A=(K[y_0,\ldots,y_d]/(F))_s$ with s the image of $\partial F/\partial y_d$ mod F. The ring A has a natural structure of Δ-ring given by

$$\delta\hat{y}_i=\hat{y}_{i+1} \qquad \text{for } 0\leq i\leq d-1$$

$$\delta\hat{y}_d=-s^{-1}(F^{\delta}(\hat{y})+\hat{y}_1(\partial F/\partial y_0)+\ldots+\hat{y}_d(\partial F/\partial y_{d-1}))$$

where $\hat{y}_j=y_j$ mod F, $\hat{y}=(\hat{y}_0,\ldots,\hat{y}_d)$ and F^{δ} is obtained from F by applying δ to all coefficients of F. Moreover $V=\operatorname{Spec} A$ becomes a Δ-variety over K and giving a solution of (2.1.2) is equivalent to giving a commutative diagram of Δ-schemes

$$\begin{array}{ccc} & & V \\ & \nearrow & \downarrow \\ \operatorname{Spec} M(R') & \longrightarrow & \operatorname{Spec} K \end{array}$$

with $R'\subset R$. Such a diagram can be viewed as an $M(R')$-point of V in the category of Δ-schemes over K.

Summarizing we may say roughly speaking that Δ-systems and their solutions lead to the following constructions (=interpretations):

Δ-systems	foliations	Δ-varieties
solutions	local integral subvarieties	"points" in the category of Δ-schemes.

(2.4) Although the two constructions (2.2) and (2.3) above are somewhat similar (the morphisms $V\longrightarrow \operatorname{Spec} K$ and $X\longrightarrow R$ should be viewed as analogs of eachother) we see that if we place ourselves into the foliation-theoretic standpoint then in general we cannot keep track of the field K and hence one cannot expect to obtain results as (1.3) or (1.6). The simplest example illustrating this situation is the example of "algebraic solutions" which shall be briefly discussed below, see also (II.3.18) and (II.3.19).

If $\varphi: X \longrightarrow R$ is an analytic map, from a manifold X to $R \subset \mathbb{C}^m$ and if we are given an m-dimensional foliation $\mathcal{F}$ on X, transverse to φ, then one usually says that "the solutions are algebraic" if and only if the leaves of $\mathcal{F}$ are finite coverings of R, see [J_1] p.215.

Now if V is a Δ-variety over K and β:Spec M(R') $\longrightarrow$ V is a Spec K-morphism of Δ-schemes then we say that β is an algebraic solution if $K(\beta)$ is algebraic over K (i.e. if the components β_j of the solution β in (2.1) are algebraic over the Δ-field generated by the coefficients of our algebraic differential equations).

Clearly the latter concept of "algebraic solution" is much stronger than the former one. Consequently for instance, Nishioka's theorem [Mtd] p.91 is much stronger than the classical Poincaré theorem [J_1] p.215 on "algebraicity of solutions" of the equation $f(y, \delta y)=0$ in the hypothesis that genus$(f=0) \geqslant 2$ and $f=0$ has no movable singularity, even if we place ourselves in the case of meromorphic coefficients. Of course our result (II.3.19) above is a generalisation of Nishioka's theorem to higher dimensions.

(2.5) Now Δ-varieties obtained in (2.3) are always affine. To get arbitrary varieties one has to glue Δ-systems in an obvious way. Similarily by glueing Δ-systems one can obtain foliations on the total space X of more complicated projections $\varphi: X \longrightarrow R$. This glueing process is very classical cf. [Poin]. Saying that "there is no movable singularity" is then equivalent to saying that we may "compactify" our data namely:

1) we may "compactify" our Δ-variety V and get a projective Δ-variety $\overline{V}$ or

2) we may "compactify" our $\varphi: X \longrightarrow R$ and get a proper map $\overline{\varphi}: \overline{X} \longrightarrow R$ and a foliation on $\overline{X}$ extending that on X which is still transverse to $\overline{\varphi}$.

In the foliation-theoretic setting there is of course Ehresmann's theorem $[J_1]$ p.210 saying that $\overline{X} \longrightarrow R$ must be isotrivial. As we have seen the differential algebraic analog is comparatively rather subtle.

(2.6) A few words about the divisor of movable singularities for foliations. Suppose $\overline{\varphi}:\overline{X} \longrightarrow R$ is a proper map of analytic manifolds dim R=1 and suppose $\mathcal{F}$ is a 1-dimensional foliation on some open set $X \subset \overline{X}$ transverse to the restriction $\varphi : X \longrightarrow R$. If $\mathcal{F}$ has "meromorphic singularities" on $\overline{X}$ then the support of these singularities will be a divisor D on $\overline{X}$. Some components of D will be contracted by $\overline{\varphi}$ to points; these points will appear as "fixed singularities of the solutions". There may be other components of D which dominate R; these will give rise to "movable singularities of the solutions" so it is natural to call the sum of these components (with certain natural multiplicities) the "divisor of movable singularities". It is this construction which stands as a model for the definition in (I.2.3).

(2.7) Using the analogy between foliations and Δ-varities we should say that if V is a Δ-model of F/K then any element of F^{Δ} should be viewed as an "algebriic prime integral" of the Δ-systems "covering" V. Along the same lines strongly normal extensions should be viewed (as explained in [NW],Appendix) as an analog of flat connections in principal bundles, having no non-trivial algebraic prime integral.

(2.8) Let's close by discussing the possibility of "reducing" Δ-varieties to foliations. Let K be the Δ-field generated over $\mathbb{C}$ by finitely many meromorphic functions $\gamma_1,\ldots,\gamma_N$ in some region $R \subset \mathbb{C}^m$ and let V be a partial Δ-variety over K. Clearly δ_V are not vector fields on V. But is the following statement true ?

(*) There exists a diagram of Δ-schemes, cartesian in the category of schemes:

$$\begin{array}{ccc} V\otimes K_a & \longrightarrow & X \\ \big\downarrow & & \big\downarrow f \\ \operatorname{Spec} K_a & \xrightarrow{\ e\ } & S \end{array}$$

with e dominant, X and S being varieties over $\mathbb{C}$ (and so δ_X, δ_S being vector fields on X and S respectively).

If (*) was true we could say that Δ-varieties are reduced in some sense to "vector fields on the total space of a family of varieties". An important feature of the theory is that (*) fails in general (in fact it fails "generically" !). It is easy to produce a counterexample to (*). Let for instance $\gamma \in M(R)$ be such that $\mathbb{C}\langle\gamma\rangle := \mathbb{C}(\gamma, d\gamma/dw, \ldots$ $\ldots, d^k\gamma/dw^k, \ldots)$ has infinite transcendence degree over $\mathbb{C}$ (this holds for "most" γ in $M(R)$) and let $\overline{V}$ be the non-singular projective model of the elliptic curve over $K := \mathbb{C}\langle\gamma\rangle$ with plane equation

$$y^2 - x(x-1)(x-\gamma) = 0$$

Extend δ_K arbitrarily to some δ_F, $F := Q(\overline{V})$, let D be the divisor of movable singularities on $\overline{V}$ and put $V = \overline{V} \setminus \operatorname{Supp} D$. Clearly V is a Δ-variety and if a diagram as in (*) would exist then two possibilities may occur:

1) $\dim S = 0$
2) $\dim S > 0$

If $\dim S = 0$, $S = \operatorname{Spec} \mathbb{C}$ hence $Q(V\otimes K_a)$ would have its j-invariant equal to that of $Q(X)$ which contradicts the fact that $\gamma \notin \mathbb{C}$. If $\dim S > 0$ then $Q(S)$ would be an intermediate Δ-field between $\mathbb{C}$ and K_a which is different from $\mathbb{C}$. It is easy to see then that we must have $\operatorname{tr.deg.} K_a/Q(S) < \infty$; we would get that $\operatorname{tr.deg.} Q(S)/\mathbb{C} = \infty$, contradiction.

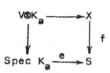

One should say that (*) holds provided $\operatorname{tr.deg.} K/\mathbb{C} < \infty$. But even in this case it will happen in general that the foliation $\mathcal{F}$ defined by the pairwise commuting vector fields δ_X has $m=\dim \mathcal{F} < \dim S$ hence the theory in [J1] Exp.5 will not apply.

Finally note that one can produce counterexamples to (*) having a quite different flavour namely with V projective over K (of course for such a V, the variety $V \otimes K_a$ is always defined over $\mathbb{C}$ by (II.1)). Indeed one just has to take $V = E \otimes_{\mathbb{C}} K$ with $K = \mathbb{C}\langle\gamma\rangle$ as in the first counterexample and E an elliptic curve over $\mathbb{C}$; then consider on V the derivation $\delta_V = \delta_K^* + \gamma\theta$ where θ is a generator of $\mathrm{Lie}_{\mathbb{C}}(E)$.

3. An example: Euler equations.

As explained in (IV.1) strongly normal Δ-function fields lead to systems which can be "linearized" by means of abelian functions and hence by θ-functions. By (III.2) this linearisation property will hold also for Δ-function fields with no movable singularity. On the other hand there is another remarkable class of systems which can be "integrated" (in a certain special sense) by θ-functions namely the algebraically completely integrable (a.c.i.) Hamiltonian systems [VM]. It is not hard to prove that in fact a.c.i. Hamiltonian systems fit into our theory of Δ-function fields with no movable singularity. Rather than discussing the abstract general situation we shall consider a very special but significant case namely the case of Euler equations (describing the movement of a rigid body with fixed gravity center [Arn]).

(3.1) Start with an ordinary Δ-field K, take $a \in K \setminus \{0\}$ and take $c_1, c_2, c_3 \in K^{\Delta}$. Now consider the purely transcendental extension $F = K(y_1, y_2, y_3)$ as a Δ-field extension of K with derivation given by:

$$\delta y_1=a(c_2-c_3)y_2y_3$$
$$\delta y_2=a(c_3-c_1)y_3y_1$$
$$\delta y_3=a(c_1-c_2)y_1y_2$$

For $K=\mathbb{R}$ viewed as a constant field this is just Euler's system [Arn].

Now $\mathbb{A}^3:=\operatorname{Spec} K[y_1,y_2,y_3]$ is a Δ-model of F/K; but note that if we embed $\mathbb{A}^3$ into $\mathbb{P}^3:=\operatorname{Proj} K[x_0,x_1,x_2,x_3]$ $(y_i=x_i/x_0$ for $1\leq i\leq 3)$ then $\partial/\partial y_i=x_0(\partial/\partial x_i)$ hence

$$\delta_F=a(c_2-c_3)y_2y_3(\partial/\partial y_1)+\ldots=$$

$$=a(c_2-c_3)(x_2/x_0)(x_3(\partial/\partial x_1))+\ldots$$

hence the divisor of movable singularity on $\mathbb{P}^3$ equals the plane at infinity $\mathbb{P}^3\setminus\mathbb{A}^3$ (with multiplicity one). In particular $\mathbb{P}^3$ is not a Δ-model of F/K.

Put now

$$\gamma_1=y_1^2+y_2^2+y_3^2 \qquad \in F$$
$$\gamma_2=c_1y_1^2+c_2y_2^2+c_3y_3^2 \in F$$

One checks immediately that $\delta\gamma_1=\delta\gamma_2=0$ (γ_1 and γ_2 are of course the classical prime integrals known from mechanics !). Put $L=K(\gamma_1,\gamma_2)$. We claim that if $(c_1-c_2)(c_2-c_3)(c_3-c_1)\neq 0$ then F/L is a Δ-function field of genus 1 with no movable singularity. To see this let V be the closed subscheme of $\mathbb{P}_L^3$ given by the equations $Q_1=0$ and $Q_2=0$ where

$$Q_1=T_1^2+T_2^2+T_3^2-\gamma_1T_0^2$$

$$Q_2=c_1T_1^2+c_2T_2^2+c_3T_3^2-\gamma_2T_0^2$$

Put $S_1=T_1/T_0$, $S_2=T_2/T_0$, $S_3=T_3/T_0$ and let L_a be an algebraic closure of L. Clearly $V\otimes_L L_a$ is an elliptic curve over L_a. We have an embedding $\mathbb{A}^3 \longrightarrow \mathbb{P}^3 \times \mathbb{A}^2$ given by

$$(y_1,y_2,y_3) \longmapsto (1:y_1:y_2:y_3, y_1^2+y_2^2+y_3^2, c_1y_1^2+c_2y_2^2+c_3y_3^2)$$

If X is the image of this embedding then V is birationally isomoRphic to $X \times_{\mathbb{A}^2} \mathrm{Spec}\, L$ $(L=Q(\mathbb{A}^2))$ showing that V is a non-singular projective model of F/L of genus 1. Let's show that $\delta(\mathcal{O}_V) \subset \mathcal{O}_V$. It is sufficient to prove that $a^{-1}\delta(\mathcal{O}_V) \subset \mathcal{O}_V$ so we may suppose $a=1$. Put $C=K^\Delta$, $M=C(\gamma_1,\gamma_2)$. Clearly $V=Y\otimes_M L$ where $Y \subset \mathbb{P}_M^3$ is given by the same equations $Q_1=Q_2=0$. It will be sufficient to prove that $\delta(\mathcal{O}_Y) \subset \mathcal{O}_Y$ in other words we may suppose that L is constant. Put

$$\lambda_1=\lambda_2=\lambda_3=1, \quad \lambda_0=-\gamma_1$$

$$\mu_1=c_1,\ \mu_2=c_2,\ \mu_3=c_3,\ \mu_0=-\gamma_2$$

so we may write $Q_1=\sum_{i=0}^{3}\lambda_i T_i^2$, $Q_2=\sum_{i=0}^{3}\mu_i T_i^2$ and

$$\delta=-(\lambda_2\mu_3-\lambda_3\mu_2)s_2s_3(\partial/\partial s_1)+\dots$$

Put for $i,j \in \{0,1,2,3\}$, $i \neq j$

$$u_{ij}=T_iT_j(T_i dT_j-T_j dT_i)(T_0T_1T_2T_3(\lambda_i\mu_j-\lambda_j\mu_i))^{-1}$$

$u_{ij} \in H^0(\mathbb{P}^3_{ij},\Omega)$ where $\mathbb{P}^3_{ij}$ is the open set of $\mathbb{P}^3_L$ given by $T_iT_j \neq 0$. It is easy to check that $u_{ij}=u_{km}$ on $V \cap \mathbb{P}^3_{ij} \cap \mathbb{P}^3_{km}$ for $i\neq j$ and $k \neq m$. Hence u_{ij} define a global 1-form on V, call it ω. Since $\langle\omega,\delta\rangle=-1$ we get that δ is a generator of the L_a-vector space of global vector fields on $V\otimes L_a$ and hence $\delta(\mathcal{O}_V) \subset \mathcal{O}_V$ by (I.2.5). We got that F/L has no movable singularity.

(3.2) If we place ourselves in the analytic case i.e. if $K \subset M(R)$ for some region $R \subset \mathbb{C}$ then we can embed L into $M(R\times\mathbb{C}^2)$ over K by sending γ_1 and γ_2 into the coordinates u_1 and u_2 on $\mathbb{C}^2$; this is a Δ-embedding if we view $M(R\times\mathbb{C}^2)$ as an ordinary Δ-field with derivation $\partial/\partial w$ (w being a coordinate on R). Now if we look for solutions (in the sense of (IV.2)) of our system in $M(R_1)$ where R_1 is a subregion of $R\times\mathbb{C}^2$ we find that they are always expressible by means of "special values of certain automorphic functions" as in (IV.1.8) and (IV.1.9). This agrees of course with the classical fact that Euler equations can be integrated by θ-functions related to elliptic curves.

(3.3) Let's discuss the case $c_1=c_2\neq c_3$. Put in this case $\eta_1=y_3$ $\eta_2=y_1^2+y_2^2$. Clearly $\delta\eta_1=\delta\eta_2=0$. Put $E=K(\eta_1,\eta_2)$. We claim that F/E is a Δ-function field of genus zero with no movable singularity (hence coming from a Riccati equation). Indeed exactly as in (3.1) a model of F/E is the closed subscheme V of $\mathbb{P}_E^3$ defined by

$$T_3-\eta_1T_0=0$$

$$T_1^2+T_2^2-\eta_2T_0^2=0$$

It is a plane conic over E with no rational point. Now if $e\in E_a$, $e^2=\eta_2$ then $F(e)=E(e,y_1,y_2)=E(e,t)$ where $t=y_2/(y_1-e)$. One immediately checks that

$$\delta t=(a/2)\eta_1(c_3-c_1)(t^2+1)$$

hence t satisfies a Riccati equation [Mtd] so $F(e)/E(e)$ has no movable singularity. By (I.2.5) F/E will have no movable singularity.

(3.4) Last case $c_1=c_2=c_3$ in Euler equations is trivial from our point of view since in this case F/K is split.

REFERENCES

[A] M.Artin, Some numerical criteria for contractibility of curves on an algebraic surface, Amer.J.Math.84(1962),485-496.

[Ang] B.Angéniol, Familles de cycles algébriques.Schéma de Chow, Lecture Notes in Math.896, Springer 1981.

[Arn] V.Arnold, Méthodes mathématiques de la mécanique classique, Mir, Moscou 1976.

[BB] A.Białynicki-Birula, On Galois theory of fields with operators, Amer.J.Math.84(1962),89-109.

[Bing] J.Bingener, Offenheit der Versalität in der Analytischen Geometrie, Math Z. 173(1980),241-281.

[BO] W.Barth, E.Oeljeklaus, Uber die Albanese Abbildung einer fasthomogenen Kähler-Mannigfaltkeit, Math. Ann. 211-1(1974),47-63.

[BPS] C.Bănică, M.Putinar, G.Schumacher, Variationen der globalen Ext in Deformationen kompakter komplexer Räume, Math. Ann. 250(1980) 135-165.

[BPV] W.Barth, C.Peters, A.Van de Ven, Compact Complex Surfaces, Springer Verlag 1984.

[Bu_1] A.Buium, Class groups and differential function fields, J.Algebra, 89,1(1984),56-64.

[Bu_2] A.Buium, Fields of definition of algebraic varieties in characteristic zero, to appear in Compositio Math.

[Bu_3] A.Buium, Corps de définition des variétés algébriques et théorie de Galois pour les corps differentiels, C.R.Acad.Sci.Paris,300 (1985),627-629.

[Bu_4] A.Buium, Ritt schemes and torsion theory, Pacific J.Math.98(1982) 281-293.

[BW] D.Burns, J.Wahl, Local contributions to global deformations of surfaces, Invent.Math. 26(1974),67-88.

[Cass] P.J.Cassidy, Differential algebraic groups. Amer.J.Math.94(1972), 891-954.

[Ch] W.L.Chow, On the projective embedding of homogenous varieties, in: Symp. in honor of S.Lefschetz, Princeton Univ. Press 1957, 122-128.

[Che] C.Chevalley, Théorie des groupes de Lie: Groupes algébriques, Hermann, Paris 1951.

[Dou] A.Douady, Le problème des modules locaux pour les espaces $\mathbb{C}$-analytiques compactes, Ann.Sci.Ec.Norm.Sup.Ser 4,7(1974),569-602.

[FGA] A.Grothendieck, Fondaments de la Géometrie Algébrique, Sem. Bourbaki 1957-1962.

[GH] P.Griffiths, J.Harris, Principles of Algebraic Geometry, John Wiley&Sons 1978.

[GS] R.Gerard, A.Sec, Feuilletages de Painlevé, Bull.Soc.Math.France, 100(1972),47-72.

[Ha] R.Hartshorne, Algebraic Geometry, Springer Verlag 1977.

[Hu] J.E.Humphreys, Linear Algebraic Groups, Springer Verlag 1975.

[Iv] B.Iversen, Linear Determinants and Applications to the Picard Scheme of a Family of Algebraic Curves, Lecture Notes in Math. 174, Springer Verlag 1970.

[J_1] J.P.Jouanolou, Equations de Pfaff Algébriques, Lecture Notes in Math. 708, Springer Verlag 1979.

[J_2] J.P.Jouanolou, Théoremes de Bertini et Applications, Progress in Math. 42, Birkhäuser 1983.

[Kol_1] E.R.Kolchin, Differential Algebra and Algebraic Groups, Academic Press, New York 1973.

[Kol_2] E.R.Kolchin, Galois theory of differential fields, Amer. J.Math. 75(1953),753-824.

[Kol_3] E.R.Kolchin, Abelian extensions of differential fields, Amer.J. Math.82(1960),779-790.

[Kol_4] E.R.Kolchin, Constrained extensions of differential fields, Adv. in Math.12(1974),141-170.

[Kol_5] E.R.Kolchin, Differential Algebraic Groups, Academic Press, New York 1985.

[Ka] I.Kaplanski, An Introduction to Differential Algebra, Hermann, Paris 1957.

[Kaw] Y.Kawamata, Characterisation of abelian varieties, Compositio Math.48(1981),253-276.

[Kei] W.Keigher, Adjunction and commonads in differential algebra, Pacific J.Math.59(1975),99-112.

[Ker] H.Kerner, Familien kompakter und holomorph-vollstandiger komplexer Räume, Math.Z.92(1966),225-233.

[Kn] D.Knutson, Algebraic Spaces, Lecture Notes in Math.203, Springer Verlag 1971.

[Koiz] S.Koizumi, The fields of moduli for polarized abelian varieties and curves, Nagoya Math.J.,48(1972),37-55.

[Kov] J-Kovacic, On the inverse problem in Galois theory of differential fields, Ann.of Math.93(1971),269-284.

[KS] K.Kodaira, D.C.Spencer, On deformations of complex analytic structures I & II, Ann.of Math. 67(1958),328-466.

[La] S.Lang, Abelian Varieties, Interscience, New York 1959.

[Li_1] D.I.Liebermann, Holomorphic vector fields and rationality,in: Group Actions and Vector Fields, Lecture Notes in Math.956, Springer Verlag 1982.

[Li_2] D.I.Liebermann, Compactness of the Chow scheme, Sém.Norguet 1976, Lecture Notes in Math. 670, Springer Verlag 1978.

[MD_1] M.Martin-Deschamps, Surfaces de type général et de type (T_2) sur un corps de fonctions, Thèse de doctorat,1976, Univ.Paris XI

[MD_2] M.Martin-Deschamps, Propriété de descente des variétés à fibré cotangent ample, Ann.Inst.Fourier 33,3(1984),39-64

[Nag] M.Nagata, Local rings, Interscience Publishers, New York, London.

[Mi] J.S.Milne, Étale Cohomology, Princeton Univ.Press 1980

[MM] T.Matsusaka, D.Mumford, Two fundamental theorems on deformations of polarized varieties, Amer.J.Math.86(1964),668-684.

[Mtd] M.Matsuda, First Order Algebraic Differential Equations, Lecture Notes in Math.804,Springer Verlag 1980.

[Mtk] T.Matsusaka, Algebraic deformations of polarized varieties, Nagoya Math.J.31(1968),185-245.

[Mtm] H.Matsumura, Integrable derivations, Nagoya Math.J.87(1982),227-247.

[Mum] D.Mumford, Geometric Invariant Theory, Springer Verlag 1965.

[Ni] M.Nishi, Some results on abelian varieties, Nat.Sc.Report Ochanomizu Univ.9,1(1958),1-12.

[NW] W.Nichols, B.Weisfeiler, Differential formal groups of J.F.Ritt, Amer.J.Math.104,5(1982),943-1005.

[Pa] P.Painlevé, Sur les équations différentielles d'ordre quelconque à points critiques fixes, C.R.Acad.Sci.Paris 130(1900),1112-1115.

[Poin] H.Poincaré, Sur un théorème de M.Fuchs, Acta Math.7(1885),1-32.

[Pomm] J.F.Pommaret, Differential Galois Theory, Gordon & Breach 1983.

[Popp] H.Popp, Moduli Theory of and Classification Theory of Algebraic Varieties, Lecture Notes in Math.620, Springer Verlag 1977.

[Ra] N.Radu, Sur la décomposition primaire des idéaux différentiels, Rév.Roum.Math.Pures et Appl.16,9(1971).

[Ray] M.Raynaud, Faisceaux amples sur les schémas en groupes et les espaces homogenes, Lecture Notes in Math.119, Springer Verlag 1970.

[Ri] J.F.Ritt, Differential Algebra, Amer.Math. Soc.Colloq.Publ.33(1950).

[Ro] M.Rosenlicht, Some basic theorems on algebraic groups, Amer.J. Math.78(1956),401-463.

[S_1] J.P.Serre, Appendix to [Wald] below

[S_2] J.P.Serre, Cohomologie Galoisienne, Springer Verlag 1965.

[Sch] H.W.Schuster, Uber die starrheit kompakter komplexer Räume, Manuscripta Math.1(1969),125-137.

[Se_1] A.Seidenberg, Derivations and integral closure, Pacific J.Math. 16(1966),167-173.

[Se_2] Differential ideals in rings of finitely generated type, Amer. J.Math.89(1967),22-42.

[SGA 3] M.Demazure, A.Grothendieck, Schémas en Groupes I, Lecture Notes in Math. 151, Springer Verlag 1970.

[Sh_1] G.Shimura, Automorphic Functions and Number Theory, Lecture Notes in Math. 54, Springer Verlag 1968.

[Sh_2] G.Shimura, On the field of rationality for abelian varieties, Nagoya Math.J.45(1972),167-178.

[SZ] S.Saks, A.Zygmund, Analytic Functions, Elsevier Publishing Company, Amsterdam-London-New York 1971.

[U] K.Ueno, Classification Theory of Algebraic Varieties and Compact Complex Spaces, Lecture Notes in Math.439, Springer Verlag 1975.

[Vie] E.Viehweg, Weak positivity and aditivity of the Kodaira dimension II: the Torelli map, in: Classification of Algebraic and Analytic Manifolds, Birkhäuser 1983, 567-584.

[VM] P.van Moerbeke, Algebraic complete integrability of Hamiltonian systems and Kac-Moody Lie algebras,in: Proceedings of the International Congress of Mathematicians 1983 Warszawa,pp.881-901, North.Holland 1984.

[Wa] J.Wahl, A cohomological characterisation of $\mathbb{P}^n$, Invent. Math. 72,2(1983),315-323.

[Wald] M.Waldschmidt, Nombres transcendents et groupes algébriques, Astérisque 69-70(1979).

[Se_3] A.Seidenberg, On analytically equivalent ideals, Publ.Math. IHES 36(1969),69-74.

Subject index.

Vol. 1062: J. Jost, Harmonic Maps Between Surfaces. X, 133 pages. 1984.

Vol. 1063: Orienting Polymers. Proceedings, 1983. Edited by J. L. Ericksen. VII, 166 pages. 1984.

Vol. 1064: Probability Measures on Groups VII. Proceedings, 1983. Edited by H. Heyer. X, 588 pages. 1984.

Vol. 1065: A. Cuyt, Padé Approximants for Operators: Theory and Applications. IX, 138 pages. 1984.

Vol. 1066: Numerical Analysis. Proceedings, 1983. Edited by D. F. Griffiths. XI, 275 pages. 1984.

Vol. 1067: Yasuo Okuyama, Absolute Summability of Fourier Series and Orthogonal Series. VI, 118 pages. 1984.

Vol. 1068: Number Theory, Noordwijkerhout 1983. Proceedings. Edited by H. Jager. V, 296 pages. 1984.

Vol. 1069: M. Kreck, Bordism of Diffeomorphisms and Related Topics. III, 144 pages. 1984.

Vol. 1070: Interpolation Spaces and Allied Topics in Analysis. Proceedings, 1983. Edited by M. Cwikel and J. Peetre. III, 239 pages. 1984.

Vol. 1071: Padé Approximation and its Applications, Bad Honnef 1983. Prodeedings. Edited by H. Werner and H. J. Bünger. VI, 264 pages. 1984.

Vol. 1072: F. Rothe, Global Solutions of Reaction-Diffusion Systems. V, 216 pages. 1984.

Vol. 1073: Graph Theory, Singapore 1983. Proceedings. Edited by K. M. Koh and H. P. Yap. XIII, 335 pages. 1984.

Vol. 1074: E. W. Stredulinsky, Weighted Inequalities and Degenerate Elliptic Partial Differential Equations. III, 143 pages. 1984.

Vol. 1075: H. Majima, Asymptotic Analysis for Integrable Connections with Irregular Singular Points. IX, 159 pages. 1984.

Vol. 1076: Infinite-Dimensional Systems. Proceedings, 1983. Edited by F. Kappel and W. Schappacher. VII, 278 pages. 1984.

Vol. 1077: Lie Group Representations III. Proceedings, 1982–1983. Edited by R. Herb, R. Johnson, R. Lipsman, J. Rosenberg. XI, 454 pages. 1984.

Vol. 1078: A. J. E. M. Janssen, P. van der Steen, Integration Theory. V, 224 pages. 1984.

Vol. 1079: W. Ruppert. Compact Semitopological Semigroups: An Intrinsic Theory. V, 260 pages. 1984

Vol. 1080: Probability Theory on Vector Spaces III. Proceedings, 1983. Edited by D. Szynal and A. Weron. V, 373 pages. 1984.

Vol. 1081: D. Benson, Modular Representation Theory: New Trends and Methods. XI, 231 pages. 1984.

Vol. 1082: C.-G. Schmidt, Arithmetik Abelscher Varietäten mit komplexer Multiplikation. X, 96 Seiten. 1984.

Vol. 1083: D. Bump, Automorphic Forms on GL (3,IR). XI, 184 pages. 1984.

Vol. 1084: D. Kletzing, Structure and Representations of Q-Groups. VI, 290 pages. 1984.

Vol. 1085: G. K. Immink, Asymptotics of Analytic Difference Equations. V, 134 pages. 1984.

Vol. 1086: Sensitivity of Functionals with Applications to Engineering Sciences. Proceedings, 1983. Edited by V. Komkov. V, 130 pages. 1984

Vol. 1087: W. Narkiewicz, Uniform Distribution of Sequences of Integers in Residue Classes. VIII, 125 pages. 1984.

Vol. 1088: A. V. Kakosyan, L. B. Klebanov, J. A. Melamed, Characterization of Distributions by the Method of Intensively Monotone Operators. X, 175 pages. 1984.

Vol. 1089: Measure Theory, Oberwolfach 1983. Proceedings. Edited by D. Kölzow and D. Maharam-Stone. XIII, 327 pages. 1984.

Vol. 1090: Differential Geometry of Submanifolds. Proceedings, 1984. Edited by K. Kenmotsu. VI, 132 pages. 1984.

Vol. 1091: Multifunctions and Integrands. Proceedings, 1983. Edited by G. Salinetti. V, 234 pages. 1984.

Vol. 1092: Complete Intersections. Seminar, 1983. Edited by S. Greco and R. Strano. VII, 299 pages. 1984.

Vol. 1093: A. Prestel, Lectures on Formally Real Fields. XI, 125 pages. 1984.

Vol. 1094: Analyse Complexe. Proceedings, 1983. Edité par E. Amar, R. Gay et Nguyen Thanh Van. IX, 184 pages. 1984.

Vol. 1095: Stochastic Analysis and Applications. Proceedings, 1983. Edited by A. Truman and D. Williams. V, 199 pages. 1984.

Vol. 1096: Théorie du Potentiel. Proceedings, 1983. Edité par G. Mokobodzki et D. Pinchon. IX, 601 pages. 1984.

Vol. 1097: R. M. Dudley, H. Kunita, F. Ledrappier, École d'Éte de Probabilités de Saint-Flour XII – 1982. Edité par P. L. Hennequin. X, 396 pages. 1984.

Vol. 1098: Groups – Korea 1983. Proceedings. Edited by A. C. Kim and B. H. Neumann. VII, 183 pages. 1984.

Vol. 1099: C. M. Ringel, Tame Algebras and Integral Quadratic Forms. XIII, 376 pages. 1984.

Vol. 1100: V. Ivrii, Precise Spectral Asymptotics for Elliptic Operators Acting in Fiberings over Manifolds with Boundary. V, 237 pages. 1984.

Vol. 1101: V. Cossart, J. Giraud, U. Orbanz, Resolution of Surface Singularities. Seminar. VII, 132 pages. 1984.

Vol. 1102: A. Verona, Stratified Mappings – Structure and Triangulability. IX, 160 pages. 1984.

Vol. 1103: Models and Sets. Proceedings, Logic Colloquium, 1983, Part I. Edited by G. H. Müller and M. M. Richter. VIII, 484 pages. 1984.

Vol. 1104: Computation and Proof Theory. Proceedings, Logic Colloquium, 1983, Part II. Edited by M. M. Richter, E. Börger, W. Oberschelp, B. Schinzel and W. Thomas. VIII, 475 pages. 1984.

Vol. 1105: Rational Approximation and Interpolation. Proceedings, 1983. Edited by P. R. Graves-Morris, E. B. Saff and R. S. Varga. XII, 528 pages. 1984.

Vol. 1106: C. T. Chong, Techniques of Admissible Recursion Theory. IX, 214 pages. 1984.

Vol. 1107: Nonlinear Analysis and Optimization. Proceedings, 1982. Edited by C. Vinti. V, 224 pages. 1984.

Vol. 1108: Global Analysis – Studies and Applications I. Edited by Yu. G. Borisovich and Yu. E. Gliklikh. V, 301 pages. 1984.

Vol. 1109: Stochastic Aspects of Classical and Quantum Systems. Proceedings, 1983. Edited by S. Albeverio, P. Combe and M. Sirugue-Collin. IX, 227 pages. 1985.

Vol. 1110: R. Jajte, Strong Limit Theorems in Non-Commutative Probability. VI, 152 pages. 1985.

Vol. 1111: Arbeitstagung Bonn 1984. Proceedings. Edited by F. Hirzebruch, J. Schwermer and S. Suter. V, 481 pages. 1985.

Vol. 1112: Products of Conjugacy Classes in Groups. Edited by Z. Arad and M. Herzog. V, 244 pages. 1985.

Vol. 1113: P. Antosik, C. Swartz, Matrix Methods in Analysis. IV, 114 pages. 1985.

Vol. 1114: Zahlentheoretische Analysis. Seminar. Herausgegeben von E. Hlawka. V, 157 Seiten. 1985.

Vol. 1115: J. Moulin Ollagnier, Ergodic Theory and Statistical Mechanics. VI, 147 pages. 1985.

Vol. 1116: S. Stolz, Hochzusammenhängende Mannigfaltigkeiten und ihre Ränder. XXIII, 134 Seiten. 1985.

Vol. 1117: D.J. Aldous, J.A. Ibragimov, J. Jacod, Ecole d'Été de Probabilités de Saint-Flour XIII – 1983. Édité par P.L. Hennequin. IX, 409 pages. 1985.

Vol. 1118: Grossissements de filtrations: exemples et applications. Seminaire, 1982/83. Edité par Th. Jeulin et M. Yor. V, 315 pages. 1985.

Vol. 1119: Recent Mathematical Methods in Dynamic Programming. Proceedings, 1984. Edited by I. Capuzzo Dolcetta, W.H. Fleming and T. Zolezzi. VI, 202 pages. 1985.

Vol. 1120: K. Jarosz, Perturbations of Banach Algebras. V, 118 pages. 1985.

Vol. 1121: Singularities and Constructive Methods for Their Treatment. Proceedings, 1983. Edited by P. Grisvard, W. Wendland and J.R. Whiteman. IX, 346 pages. 1985.

Vol. 1122: Number Theory. Proceedings, 1984. Edited by K. Alladi. VII, 217 pages. 1985.

Vol. 1123: Séminaire de Probabilités XIX 1983/84. Proceedings. Edité par J. Azéma et M. Yor. IV, 504 pages. 1985.

Vol. 1124: Algebraic Geometry, Sitges (Barcelona) 1983. Proceedings. Edited by E. Casas-Alvero, G.E. Welters and S. Xambó-Descamps. XI, 416 pages. 1985.

Vol. 1125: Dynamical Systems and Bifurcations. Proceedings, 1984. Edited by B.L.J. Braaksma, H.W. Broer and F. Takens. V, 129 pages. 1985.

Vol. 1126: Algebraic and Geometric Topology. Proceedings, 1983. Edited by A. Ranicki, N. Levitt and F. Quinn. V, 423 pages. 1985.

Vol. 1127: Numerical Methods in Fluid Dynamics. Seminar. Edited by F. Brezzi, VII, 333 pages. 1985.

Vol. 1128: J. Elschner, Singular Ordinary Differential Operators and Pseudodifferential Equations. 200 pages. 1985.

Vol. 1129: Numerical Analysis, Lancaster 1984. Proceedings. Edited by P.R. Turner. XIV, 179 pages. 1985.

Vol. 1130: Methods in Mathematical Logic. Proceedings, 1983. Edited by C.A. Di Prisco. VII, 407 pages. 1985.

Vol. 1131: K. Sundaresan, S. Swaminathan, Geometry and Nonlinear Analysis in Banach Spaces. III, 116 pages. 1985.

Vol. 1132: Operator Algebras and their Connections with Topology and Ergodic Theory. Proceedings, 1983. Edited by H. Araki, C.C. Moore, Ş. Strătilă and C. Voiculescu. VI, 594 pages. 1985.

Vol. 1133: K.C. Kiwiel, Methods of Descent for Nondifferentiable Optimization. VI, 362 pages. 1985.

Vol. 1134: G.P. Galdi, S. Rionero, Weighted Energy Methods in Fluid Dynamics and Elasticity. VII, 126 pages. 1985.

Vol. 1135: Number Theory, New York 1983–84. Seminar. Edited by D.V. Chudnovsky, G.V. Chudnovsky, H. Cohn and M.B. Nathanson. V, 283 pages. 1985.

Vol. 1136: Quantum Probability and Applications II. Proceedings, 1984. Edited by L. Accardi and W. von Waldenfels. VI, 534 pages. 1985.

Vol. 1137: Xiao G., Surfaces fibrées en courbes de genre deux. IX, 103 pages. 1985.

Vol. 1138: A. Ocneanu, Actions of Discrete Amenable Groups on von Neumann Algebras. V, 115 pages. 1985.

Vol. 1139: Differential Geometric Methods in Mathematical Physics. Proceedings, 1983. Edited by H. D. Doebner and J. D. Hennig. VI, 337 pages. 1985.

Vol. 1140: S. Donkin, Rational Representations of Algebraic Groups. VII, 254 pages. 1985.

Vol. 1141: Recursion Theory Week. Proceedings, 1984. Edited by H.-D. Ebbinghaus, G.H. Müller and G.E. Sacks. IX, 418 pages. 1985.

Vol. 1142: Orders and their Applications. Proceedings, 1984. Edited by I. Reiner and K. W. Roggenkamp. X, 306 pages. 1985.

Vol. 1143: A. Krieg, Modular Forms on Half-Spaces of Quaternions. XIII, 203 pages. 1985.

Vol. 1144: Knot Theory and Manifolds. Proceedings, 1983. Edited by D. Rolfsen. V, 163 pages. 1985.

Vol. 1145: G. Winkler, Choquet Order and Simplices. VI, 143 pages. 1985.

Vol. 1146: Séminaire d'Algèbre Paul Dubreil et Marie-Paule Malliavin. Proceedings, 1983–1984. Edité par M.-P. Malliavin. IV, 420 pages. 1985.

Vol. 1147: M. Wschebor, Surfaces Aléatoires. VII, 111 pages. 1985.

Vol. 1148: Mark A. Kon, Probability Distributions in Quantum Statistical Mechanics. V, 121 pages. 1985.

Vol. 1149: Universal Algebra and Lattice Theory. Proceedings, 1984. Edited by S. D. Comer. VI, 282 pages. 1985.

Vol. 1150: B. Kawohl, Rearrangements and Convexity of Level Sets in PDE. V, 136 pages. 1985.

Vol 1151: Ordinary and Partial Differential Equations. Proceedings, 1984. Edited by B.D. Sleeman and R.J. Jarvis. XIV, 357 pages. 1985.

Vol. 1152: H. Widom, Asymptotic Expansions for Pseudodifferential Operators on Bounded Domains. V, 150 pages. 1985.

Vol. 1153: Probability in Banach Spaces V. Proceedings, 1984. Edited by A. Beck, R. Dudley, M. Hahn, J. Kuelbs and M. Marcus. VI, 457 pages. 1985.

Vol. 1154: D.S. Naidu, A.K. Rao, Singular Pertubation Analysis of Discrete Control Systems. IX, 195 pages. 1985.

Vol. 1155: Stability Problems for Stochastic Models. Proceedings, 1984. Edited by V.V. Kalashnikov and V.M. Zolotarev. VI, 447 pages. 1985.

Vol. 1156: Global Differential Geometry and Global Analysis 1984. Proceedings, 1984. Edited by D. Ferus, R.B. Gardner, S. Helgason and U. Simon. V, 339 pages. 1985.

Vol. 1157: H. Levine, Classifying Immersions into $\mathbb{R}^4$ over Stable Maps of 3-Manifolds into $\mathbb{R}^2$. V, 163 pages. 1985.

Vol. 1158: Stochastic Processes – Mathematics and Physics. Proceedings, 1984. Edited by S. Albeverio, Ph. Blanchard and L. Streit. VI, 230 pages. 1986.

Vol. 1159: Schrödinger Operators, Como 1984. Seminar. Edited by S. Graffi. VIII, 272 pages. 1986.

Vol. 1160: J.-C. van der Meer, The Hamiltonian Hopf Bifurcation. VI, 115 pages. 1985.

Vol. 1161: Harmonic Mappings and Minimal Immersions, Montecatini 1984. Seminar. Edited by E. Giusti. VII, 285 pages. 1985.

Vol. 1162: S.J.L. van Eijndhoven, J. de Graaf, Trajectory Spaces, Generalized Functions and Unbounded Operators. IV, 272 pages. 1985.

Vol. 1163: Iteration Theory and its Functional Equations. Proceedings, 1984. Edited by R. Liedl, L. Reich and Gy. Targonski. VIII, 231 pages. 1985.

Vol. 1164: M. Meschiari, J.H. Rawnsley, S. Salamon, Geometry Seminar "Luigi Bianchi" II – 1984. Edited by E. Vesentini. VI, 224 pages. 1985.

Vol. 1165: Seminar on Deformations. Proceedings, 1982/84. Edited by J. Ławrynowicz. IX, 331 pages. 1985.

Vol. 1166: Banach Spaces. Proceedings, 1984. Edited by N. Kalton and E. Saab. VI, 199 pages. 1985.

Vol. 1167: Geometry and Topology. Proceedings, 1983–84. Edited by J. Alexander and J. Harer. VI, 292 pages. 1985.

Vol. 1168: S.S. Agaian, Hadamard Matrices and their Applications. III, 227 pages. 1985.

Vol. 1169: W.A. Light, E.W. Cheney, Approximation Theory in Tensor Product Spaces. VII, 157 pages. 1985.

Vol. 1170: B.S. Thomson, Real Functions. VII, 229 pages. 1985.

Vol. 1171: Polynômes Orthogonaux et Applications. Proceedings, 1984. Edité par C. Brezinski, A. Draux, A.P. Magnus, P. Maroni et A. Ronveaux. XXXVII, 584 pages. 1985.

Vol. 1172: Algebraic Topology, Göttingen 1984. Proceedings. Edited by L. Smith. VI, 209 pages. 1985.